Praise for *My Father and Other Animals*

'A farmer, his son ... what to forget and what to remember, how to inherit and when to change. Sam Vincent has learned a few things from his dad, and among them is how to pay attention. For those of us who hunt our food plastic-wrapped in aisle three, this is a funny, kindly and observant introduction to the landscape that wraps around our cities, keeps us alive and means the world.' —Kate Holden

'*My Father and Other Animals* is a beautiful tale of legacy, family, and a millennial finding his place in the world.'
—*The Weekend Australian*

'A witty memoir of taking on the family farm [that] reckons with Indigenous dispossession and climate change'
—*The Conversation*

'Vincent paints a funny and familiar picture of the generational differences between father and son. The handyman father who can't be slowed even by doctors' orders, and the son who can hardly be described as anything resembling "handy".'
—Grattan Institute, 2022 Prime Minister's Summer Reading List

'Sam Vincent's memoir *My Father and Other Animals* complicates clichéd notions of rural life ... a wholesome book with some serious arguments to make about sustainability.'
—*Sydney Review of Books*

'I had tears well up at some points and laughed out loud at others. I cheered Sam's fig orchard on and sympathised with him, his parents and his sisters during the difficult family discussions.' —*Australian Rural & Regional News*

'A window onto the new rural Austral[illegible]'
—*The Sydney Morning Herald*

'[Vincent's] memoir recounts, with humour and warmth, the many ups and downs that ensued as he gradually came to think differently not only about farming, but also his relationship with his father.' —*The West Australian*

'[Sam Vincent] is a careful observer and fine writer who can do justice to complex issues, and this book is a cracker.' —*The Daily Telegraph*

'A story of how a son learns about his own family just as much as how he learns to become a farmer' —*The Guardian*

'For any reader desiring to understand contemporary rural Australia, this entertaining and important book is a must-read … Refreshingly honest, acutely observed, witty, lyrical and delightfully ironic, *My Father and Other Animals* is a pure delight.' —Charles Massy

'At once a rollicking comic memoir and a thoughtful meditation on Australian farming, *My Father and Other Animals* is a warm, surprising and beautifully crafted book.' —Billy Griffiths

'Not only is *My Father and Other Animals* beautifully written, but its insights are deep and compelling. Combining hilarious comic timing with brutal brave honesty, it is an effortless journey into complex ideas and histories, and an introduction to what I think may be one of the most hopeful stories today – that of regenerative agriculture and managing livestock to restore deeply degraded land.' —Anna Krien

# MY FATHER AND OTHER ANIMALS

# MY FATHER AND OTHER ANIMALS

## HOW I TOOK ON THE FAMILY FARM

SAM VINCENT

Published by Black Inc.,
an imprint of Schwartz Books Pty Ltd
Wurundjeri Country
22–24 Northumberland Street
Collingwood VIC 3066, Australia
enquiries@blackincbooks.com
www.blackincbooks.com


First published in 2022; this edition published in 2023

Parts of this book were previously published in *Griffith Review* and *The Monthly*.
The names of some individuals have been changed.



9781760644840 (paperback)
9781743822623 (ebook)

A catalogue record for this book is available from the National Library of Australia

Cover design by Akiko Chan
Text design and typesetting by Tristan Main
Cover image by Getty/221A
Author photo by Lean Timms
Map by Alan Laver
Photo on page 291 by Lauren Carroll Harris

Printed in Australia by McPherson's Printing Group.

Supported by

This project has been assisted by the Australian Government through the Australia Council, its arts funding and advisory body, and by artsACT.

*To my parents, for this time of gifts.*

***Gollion as it is today***

Key:

1. Rear End
2. Sam's Crossing
3. Derrawa Dhaura (formerly Bald Hill West)
4. Scar Tree Dogleg (formerly Bald Hill East)
5. Billabong
6. Scribbly Gum
7. Kungsladen
8. Willow
9. Powerline Track
10. Keyline
11. Westaway
12. Hy-fer
13. Carey Lane
14. Dead Horse Yellow Box
15. Dead Horse
16. Swamp
17. Caesar South
18. Caesar North
19. Hay Lower
20. Hay Upper
21. Ablett
22. Coleman East
23. Coleman West
24. Woolshed
25. Fernleigh House
26. Old Nut Orchard
27. Driveway Paddock
28. Cattle Yard
29. Ram
30. Siphon
31. Solar
32. Tree of Heaven
33. Billy's Coat Hanger
34. Chain of Ponds
35. Darmody
36. Top Orchard
37. Bottom Orchard (including fig orchard)
38. Murrumbateman Creek
39 House and gardens

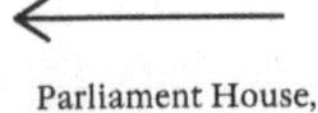

Parliament House, Canberra, 21 km

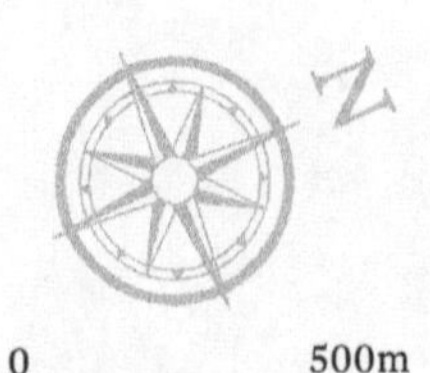


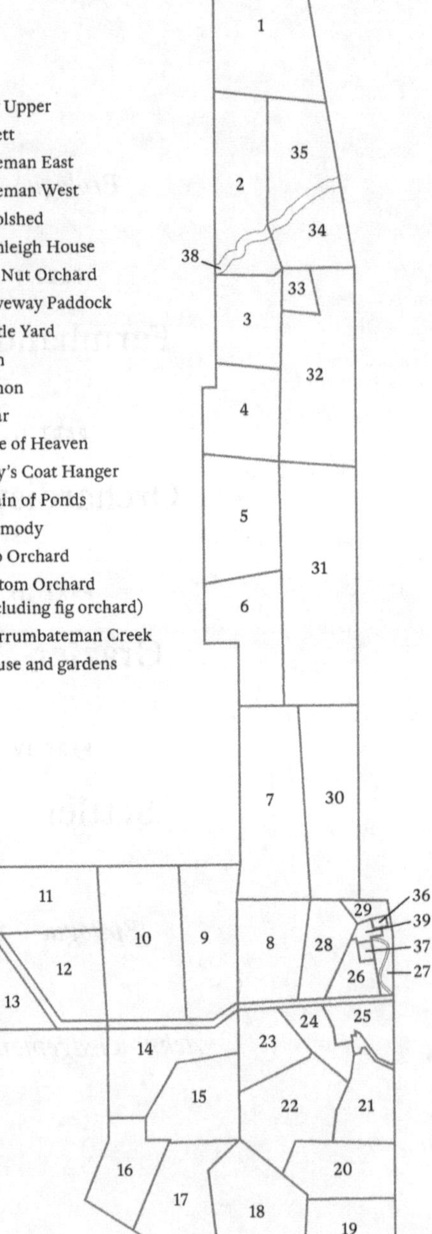

A few years ago, I was messing about at home one Sunday when my mother called and told me my father was in an ambulance after sticking his hand down a woodchipper, but that I needn't worry because 'it wasn't like that scene in *Fargo* or anything.' *Fargo*, the Coen brothers' 1996 black comedy/crime thriller, was a favourite of my parents. I knew the scene alright: Frances McDormand catching Peter Stormare in the act of mulching Steve Buscemi; the chipper's engine whining from the effort of a still-socked human leg; the snowy bank of a Minnesota lake repainted red ...

So, Dad hadn't been dismembered – a consolation, undoubtedly. But with my father, you could never be sure. Another weekend, in the winter of 1990 (I remember the date because I was in kindergarten and recounted the tale at show and tell), he cut off much of his left big toe with a chainsaw – then drove himself to hospital. He had first called in to his GP, who, upon seeing a patient hop unannounced past reception with a foot wrapped in an engine rag and God-knows-what wrapped in a handkerchief, turned pale and declared the job beyond his mandate. It wasn't until hours later that Dad rang us from hospital, the surgery a success. The sole clues to his absence were the chainsaw and earmuffs left uncharacteristically in the paddock, and the upturned boot beside them, which my sisters and I dared not look inside.

This time, the only clean-up required of my mother was to wheel the abandoned woodchipper back into the shed before night fell. My father had been chipping tree branches – shoving them all in at once rather than feeding them in separately – when the chute had blocked. Then, ignoring all three warnings prominently displayed on the machine's side – HEARING AND EYE PROTECTION MUST BE WORN; KEEP HANDS AND FACE AWAY FROM INLET OPENING; USE WOODEN STICK ONLY TO CLEAR BLOCKED CHUTE – he had picked up a length of metal pipe and thrust it inside.

He must have deduced that he couldn't drive himself to hospital with a mangled hand (at the time he drove a manual transmission; had it been the automatic he drives today he might have given it a shot), so he appeared at the kitchen window, where my mother was reading, and requested she call an ambulance. My parents' farm is forty-five minutes from Canberra, the last ten on corrugated dirt roads. When the paramedics arrived, trailing a plume of dust, they asked him to rate the pain on a level of 0 to 10; my father replied 'five', which the paramedics took to mean the pain was bearable, but to anyone who knew my father meant he was in a great deal of pain indeed.

I drove out to the farm the next day. It was late winter, sunny but brisk, and my father was dressed in his characteristic short shorts and motheaten jumper. He was in good spirits: his right thumb had been broken in three places – 'shattered' was the word he used with detectable pride – but the rest of his hand was just badly bruised. He seemed more annoyed that the accident had stopped him making compost – the acacia and tagasaste trees he'd been chipping being the first ingredients – which he customarily brewed into a 'tea' each year and sprayed on his pasture as fertiliser. His cattle were calving; the idea was that by the time the newborn calves were ready to start eating grass in spring, the pasture would be 'turbocharged' by compost tea. He spoke of the physiotherapy exercises he would do, and his confidence that 'in no time' his thumb would make a near full recovery.

But when my mother made coffee for the three of us, I sensed an unspoken feeling of *Here we are again*. Farming families often need a crisis to start a conversation about succession, but what to choose from a medical file as voluminous as David Peter Vincent's? There had, in the previous five years, been a broken wrist (sustained while trying to untangle a rope from a heifer); two cracked ribs (learning to surf in his sixties); a bulging hernia (his 'ham sandwich', tucked into his abdomen when he worked on the farm by a huge belt of the kind won at prize fights, albeit less ornate); and numerous 'angry' skin cancers (removed by his exasperated dermatologist only to be replaced by new ones, cultivated – *farmed*, even – thanks to my father's precocious taste for distressed clothing). And now this. Just that morning, Mum had caught him sneaking up a ladder to patch the net covering the top orchard with his 'good' hand. This was his interpretation of the doctor's orders to 'take it easy'.

The average age of an Australian farmer is fifty-eight; my father was sixty-six. I realised, warming my hands on my mug of coffee, that his next accident was an inevitability, and that when it happened, my mother might have to make another, harder, phone call. Dad had recently taken to telling her that if he wasn't back at the house in time for dinner, she should check that he wasn't dead under a tractor somewhere; that morning, when she pleaded with him not to climb ladders in his state, he'd cheerfully agreed that 'most farming careers are ended by falls from ladders.' These were meant as jokes, but now they seemed like premonitions.

Succession is a dirty word in any industry, but farming seems particularly averse: if mentioned at all it is in slippery euphemisms – being *put out to pasture*; *buying the farm*. 'Life on the land' means just that: if your job is your life, retirement can mean death. There was no need for Dad to stop farming just yet, and if he or Mum were thinking about it, they didn't say so. But getting him to heed his doctor and take it easy would be a challenge. This is a man who, on childhood beach holidays, pushed wheelbarrows filled with seaweed – 'cattle just love the stuff' – through throngs of suntanners back to our family car (he may be the only person in the state of New South Wales to know that you are legally allowed to forage twenty kilograms per person per day); not for him a summer book and a piña colada. He needed 'projects' – mechanical, architectural, agricultural – to relax. Always active, my father was like a shark: stop him moving, and he would suffocate.

And besides, in his eyes, he already had a retirement project – that's what farming was for.

Farming runs through my family like the tenderloin down a steer's back. My late maternal grandfather, whom I called Papa, was a patrician grazier, the son of a Melbourne cattle baron. He was a long-serving president of the Royal Agricultural Society of Victoria and lived a squatter's life of horseback mustering, lawn-tennis parties and gentlemen's clubs. (My father tells a story of once answering Papa's phone and taking a message from an acquaintance, a 'Mr Fraser from Hamilton', a.k.a. the prime minister.) But my uncles didn't want to take over Papa's farm, and my mother and my aunt – who still checks the price of weaner calves and fat lambs each week despite living in Byron Bay – weren't asked. It was sold before I was born.

From there, the farming line changed sides. My parents bought their farm, then two hundred overgrazed acres, in 1983, the year before I was born. Dad, the clever son of a clever man – my paternal grandfather, born in a bark hut in the Victorian bush and forced to leave school at age twelve by the Depression – was never going to become anything other than a white-collar professional. But although he co-founded a successful economics consultancy and is proud of his role in helping to liberalise the Australian economy, he has always looked uncomfortable in a suit. Among my first memories are the filthy orange coveralls he wore to check the sheep in early each morning, only to transform, like a bald Clark Kent with dirt under his fingernails, into a tie and dress shirt for the drive to the office. Returning home one night, still in his Canberra clothes, he shot a fox in my sister Lucy's bedroom while the rest of us were sitting down to dinner (roast chicken, which we figure attracted it inside).

Dad was raised in Melbourne but spent his school holidays helping out on a cousin's farm a few hours to the north-west ('driving trucks loaded with ten feet of teetering hay, years before I got my licence'); Mum, whom he met studying agricultural science at university, pined for a return to the land. They named their block 'Gollion', after a Swiss village where, while backpacking in the 1970s, they had found relations of an ancestor of Dad's, the first Vincent to arrive in Australia, in 1854.

Gollion was once part of Fernleigh, a farm established in the 1870s, but it is my parents who have most shaped its landscape since white settlement. Armed with the boomer audacity that made the postwar world seem theirs to conquer, Mum and Dad built themselves a house and set about constructing their Eden. Dams were bulldozed, fences strained, and my three sisters and I enlisted to plant thousands of trees on freezing winter days.

It was a process of unlearning. My parents graduated from university in the wake of the 'green' revolution, the twenty-year period of intensive technological changes in the global food system following the shortages of World War II. Their teachers advocated the use of synthetic fertilisers, pesticides and herbicides, modern machinery to efficiently sow and harvest, and the planting of hybridised monocultures to increase yield size. All that mattered was productivity. This was the seed of modern factory farming; its emphasis on agriculture as a means of production divorced from the rhythms of nature was eventually extended to cattle, hens and pigs as much as to wheat, corn and rice.

But my parents also graduated in the wake of the countercultural movements of the 1960s and '70s, when Rachel

Carson's *Silent Spring* was required reading in the communes of California, and a compost toilet was all you needed to fertilise your homegrown crop of choice. My mother especially, who worked as a high school science teacher after university, became interested in permaculture, an alternative agriculture based on a new buzzword – 'sustainability' – which my father, with his economist's brain, likes to summarise as meaning 'there *is* such thing as a free lunch.'

At Gollion, Mum and Dad subscribed to *Grass Roots* (the self-sufficiency bible of 1980s Australia), grew vegetables, kept chickens and established orchards. Over the years, my parents became less conventional in their methods. Confounding the old-school graziers around them, they brewed compost tea to fertilise their pastures organically, restored their section of a creek to the 'chain of ponds' it would have resembled before land clearing turned it into a drain, and introduced a holistic management regime in which livestock is moved regularly through many small paddocks, replicating the grazing patterns of wild herd and predator ecosystems.

They believed farming should be conducted within the environment, not against it. For the first time in a century, Gollion's paddocks remained unploughed, because ploughing destroys soil structure and releases carbon into the atmosphere. They weren't interested in growing high-yield, herbicide-and-pesticide-dependent monocultures: through trial and error, they saw for themselves how depleting soils, poisoning wildlife and reducing biodiversity caused more problems than they solved.

As they gradually reduced their inputs to only what nature could provide – sun, rain, compost, dung – their outputs continued to rise. Gollion more than tripled in size with the acquisition of neighbouring parcels of land; the acres, in my father's words, 'bred like rabbits', as if this too were an organic process.

But here's the thing: Dad, who 'retired' in 2003, always saw Gollion as a hobby farm. Some people, he reasoned, play golf when they retire; he managed 150 cows and their calves, two large orchards, a small mob of sheep and a shanty-town of sheds. While his former colleagues cruised the South Pacific, Dad sought adventure atop dodgy ladders or by roll-starting a rollbar-less tractor. ('Best not tell your mum I've been doing that again.')

Gollion was never meant to be a legacy project, and I was never raised to be its successor. If anything, I was raised to believe that farmers – as opposed to the 'go-getters' (a favourite expression of Dad's) of the service economy – were largely 'losers' (another favourite): either lazy aristocrats who inherited the good fortune of their fore-bears, or uncouth bumpkins with their hands out for government assistance.

A common lament of Australian policymakers is that, with emerging markets to our north willing to pay record prices for our produce, Australia could be the food bowl of Asia – if only we could hold onto our farmers. In one of the most urbanised countries on Earth, the drain from coun-try to city has become a torrent, leaving in its wake broken communities and farm gates locked for good. But Dad showed less interest in my continuing what he established than in my leading my 'own life' (that is, procreating and buying a house). He didn't get the memo that most millen-nials are renters, unmarried and haven't had kids yet. Nor the one that you're only able to combine full-time farming with full-time city work if you opt out of third-wave feminism. That's what happens when you spend much of your time in the back paddock with no one for company but Suey the sheepdog.

And so, although I spent much of my childhood building hideouts among the gum trees and skinny dipping in a choice of thirty dams, I didn't know how to join two pieces of wire with a figure-eight knot or tell the difference between wallaby grass and kangaroo grass; I'd never put a rubber ring around a bull calf's testes or pruned a cherry tree. My sisters had moved interstate as soon as they finished high school; I lived at the farm a few years longer and now lived within visiting distance, in Canberra, but that's all the farm was to me – the setting for my parental visits. Mum pottering in her garden and Dad working in his paddocks: it was something that had always been, and always, I thought, would be.

But that winter's day, drinking coffee with them, the stakes suddenly seemed very high. We had never been close, my father and I. Australian rules football, our only shared interest, provided fodder for conversation during autumn and winter, but that meagre fare was quickly exhausted ('Did you see the game last night?' 'Yeah, good, wasn't it?'); in summer, I would try to sympathise with him about dry spells in the weather, but talking about droughts with a farmer as chitchat is cruel – like pressing an ex-con starting a new life outside what they were in for.

What did I know of this man, really? His motivations, desires, fears and disappointments – let alone the skills and wisdom he had gained over three decades of heterodox farming? I had never asked what farming meant to him; when he spoke of the compost that would 'turbocharge' his pasture, I just nodded and smiled. He had told me once that if he had his 'time again' he would have been an orchardist instead of an economist. And that's a shame, don't you think?

Although he was too polite to say so, I suspect he thought I was a 'loser', or at least well on the way. At twenty-nine, I was single, childless, casually employed in a dead-end office job and struggling to sell the odd piece of freelance journalism. If 'millennials' only exist in opposition to 'baby boomers', the fact that my father was able to buy a 'hobby' farm at roughly the same age as I was struggling to pay the rent seemed to confirm the dichotomy.

The beneficiary of free tertiary education, affordable housing, a soaring labour market and an unbridled faith in the social democratic project, he had risen above his station to join the middle class and heaped opportunities upon his family when he got there. Dad couldn't understand why I, too, couldn't 'get ahead'. He expressed this in snide remarks when I would ask to borrow his car because mine had broken down, or when I showed up to family Christmas alone, again.

I had the time to spend a couple of days working at Gollion each week, and nothing better to do. In helping to protect my father from himself, I figured I might begin to know him – and he, me. And so, with all the deliberation it takes to decide that you're having pizza for dinner or that you'd better take a jacket with you as you head out the door, before I'd finished my coffee that morning, I'd given myself a vocation.

That's what I would do – become my father's farmhand.

PART I

# Farmhand

When farmers are taught, starting in childhood, by parents and grandparents and neighbors, their education comes 'naturally,' and at little cost to the land. A good farmer is one who brings competent knowledge, work wisdom, and a locally adapted agrarian culture to a particular farm that has been lovingly studied and learned over a number of years. We are not talking here about 'job training' but rather about the lifelong education of an artist, the wisdom that come from unceasing attention and practice.

**Wendell Berry**

# 1

He said yes. Or, more accurately, Mum said yes. Dad said, 'I always do what my wife tells me.'

I was to report for work the following Wednesday, but not in my 'Sunday best'. It was a curious piece of advice. My father was a non-churchgoer but had been raised a Methodist. The heaviest piece of baggage he'd carried into adulthood was a work ethic that never allowed for the wearing of Sunday best, because every day was a workday. (My mother cursed the denomination rather than the man when he spent our 'family time' gathering seaweed or building sheds. She was of Anglican stock and had no trouble spending her holidays on holiday.) In the years since his retirement, my father had prided himself on not having worn a tie. But that didn't mean he'd stopped wearing a uniform.

We didn't know where they came from, but under his tatty farm sweaters he'd taken to wearing second-hand T-shirts of various degrees of poor taste; he would plead ignorance to their meaning until educated by his children. One featured a cartoon of 'tank man', the unidentified Chinese dissident photographed staring down a column of

People's Liberation Army tanks in Tiananmen Square in 1989. There was a caption: 'Chinese Olympic Shooting Team Practice.' Another depicted three felons in prison fatigues, their eyes redacted with black bars and above them some kind of grammarless threat: 'EVERYTHING YOU HATE HERE'S THE MILITIA'. For my first day working on the farm, I opted for a plaid shirt and old Levis that had been washed colourless. When I found my father in his tool shed, at 8 a.m., he was wearing a black T-shirt emblazoned with the red star of socialism on the back, and on the front the Cyrillic letters 'ЖГБ'. A blonde Russian temptress leant forwards suggestively on his chest and whispered, through pursed lips, '... Still watching you'. It was 20 August 2014. We were ready to farm.

We spent the day repairing a broken fence between two paddocks. I was eager but nervous, and embarrassed about feeling nervous, this being my dad and my childhood home after all, even if I was seeing them from a new perspective.

My father, too, was playing an unfamiliar role: teacher. He had loaded the tools we would need into the back of his ancient ute ('hasn't been on a road in twenty years but she starts every time'). I drove us to our job and brought the vehicle to a stop where he directed, and then he conducted a stocktake.

'These are your *steelies*' (black steel fenceposts, holes drilled in each so that wires could pass through); 'this is your *donger*' (the manual driver for ramming them in – also a slang word for penis among Australian schoolboys of my father's generation). The wire strainers (two unconnected metal jaws to be fastened around two strands of wire before ratcheting them together) – I had seen him use these before, albeit while straining calves out of their mothers' birth canals, a practice that appalled veterinarians. The 'plain' wire was coiled around a large metal spinner that took the two of us to lift out of the ute. The 'barb' was wrapped around a much lighter wooden spinner; barbed wire comes in smaller coils and is unwound by slowly walking backwards, the wire extending in front of you like the cable of a Looney Tunes TNT detonator.

'Building a fence is like building a house,' Dad said, as if I knew how to do that. 'There are steps to follow to make it secure.'

*One.* Plant the corner posts, which support the wires and steelies between them and are braced for strength. At Gollion, corner posts are hewn from hardwood logs. Dad had erected the corner posts for this fence years earlier; the steelies between them, low-quality 'Chinese junk', had snapped under pressure, leaving behind jumbles of wire that we would reuse in repairing the fence. Before putting the corner posts into their three-foot-deep holes, he told me, it was important to paint their ends with sump oil to stop termites from eating them.

*Two.* Strain two lengths of plain wire between the corner posts, and use this as a guide to place the new, Australian-made steelies in a straight line. This temporary wire was medium tensile, meaning it could be manipulated by hand and tied using the only knot I knew: a granny knot. 'A bit girly,' Dad said upon inspecting my handiwork, 'but it'll do.' He then used his good hand to join the wires together in the middle of the fence, feeding their ends into the strainers before cranking the chain. The newly taut wires sprang off the ground into a waist-high straight line.

Ramming in the steelies (*three*) was hard, loud work: lift the donger by its handles; ease its lip over the steelie; then RAM, RAM, RAM, metal clanging on metal. A donger resembles a World War II bazooka, the kind toy soldiers get down on one knee to fire. After ramming in a dozen steelies, my arms ached. I had over a hundred to go.

'There's a petrol-engine post-driver somewhere in the shed,' Dad mentioned casually, an afterthought.

I could be getting a *machine* to do this?

'That'd be cheating.' He grinned. The bastard.

On the ten-minute drive back to my parents' house for morning tea, I noticed fences in disrepair wherever I looked. Some had collapsed under the weight of fallen tree branches. Others had been pushed over by a hungry cow or thirsty bull. I did a calculation in my head. We had thirty-five paddocks, all roughly the size of the one we had been working in, which had a kilometre of fences. Subtracting the fences that double as boundaries between two paddocks, and adding the ones around tree plantations, I estimated we had forty kilometres of fences. I ran my arithmetic past Dad.

'Just imagine what it's like for *real* farmers,' he replied. 'They've got thousands of kilometres of fence-lines to maintain. At that scale it's like trying to maintain a border between two countries.'

Why in Australia are farms measured in Belgiums? Irelands, Israels and Switzerlands are also in circulation, but farmers are notoriously set in their ways. To say that one's landholding is 'bigger than Belgium' (give or take

30,688 square kilometres, 7.58 million acres or 3.06 million hectares) is the preferred boast.

When Clifton Hills, a cattle station on the edge of the Simpson Desert, was listed for sale in 2018, the simple sales pitch was: 'It's more than half as big as Belgium!' When Anna Creek, the 'neighbouring' farm (their homesteads are hundreds of kilometres apart), sold in 2016, the Adelaide *Advertiser* described it as 'roughly' the size of Belgium.

Tieyon, in the geographic centre of Australia, is my father's idea of a 'real' farmer's farm. I don't remember Dad's connection – he'd done some work for the owner, or was a friend of a friend – but in the winter of 1997, during a family road-trip, we stayed the night. At twelve years old I wasn't interested in farming, but I *was* interested in Tintin, and when Dad said the farm we were visiting was 'as big as Belgium', I pictured the cartoon-strip reporter rushing out the door of his Brussels apartment, his white fox terrier, Snowy, at his heels.

Tieyon had many dogs, but none of them looked like Snowy. I remember them as mongrels, chained up and snarling, each housed in a kind of large pigeonhole. Tieyon also had many camels, feral descendants of caravans used in the days before road and rail reached the desert. And Tieyon had many cattle – 10,000 in a good year, when the autumn rains fill what are usually 'lakes' in name only, and the sand dunes erupt with wildflowers.

The only other thing I remember about Tieyon is that the whole time we were there, my father never once mentioned to anyone that we, too, lived on a farm. Was he embarrassed?

Gollion is a microstate in the atlas of Australian farms. It is skinny – less than 500 metres wide in some parts – and just over four kilometres long. Ask my father what kind of soil it has (the third question farmers ask one another after farm size and annual rainfall), and he will say, 'Crap', but its geology is as diverse as its geography.

A bird flying west from Lake George, north of Canberra, will enter Gollion airspace at the edge of the Yass Valley, where the Australian Capital Territory's northern tip pokes into New South Wales. The land here is flat and swampy, with rich alluvial soil and thickets of long grass that is wet for much of the year and harbours snakes when it isn't. Continuing its journey, the bird will pass the original Fernleigh – the brown rusted iron roofs of scattered sheds, surrounded by orderly fields where wheat was once grown. Today the land sways with a mixture of European grasses and is dotted with yellow box and stringybark trees, some of which survived the early settlers' enthusiasm for ringbarking. Our small flock of dorper sheep enjoys their shade. Flying on, the bird will climb with the landscape and spot my parents' house and garden, tucked out of the wind on the side of a hill. (If the bird is a parrot, it will now descend to inspect the orchards near the house; in spring there will be cherries to plunder; in summer, stone fruit; in autumn, apples, pears, quinces and figs.)

The bird's shadow expands and contracts now, contorting with each hill it passes, flying at an altitude of 800 metres. If the bird is a pollinator, it might stop here to sip the nectar from a plantation of tagasaste trees; if it's a bird of prey, it might snatch a rabbit returning to its warren. This is what Dad calls 'hungry country', and the soil here is shallow and rocky. Should the bird perform a brief loop at this point, it may be surprised at its ascent: the valley floor has been left far below, and if it's a winter's morning, so too has the fog.

For the next two kilometres, the bird will cruise across a floodplain – green or Hessian bag brown depending on the season, interrupted by the small dark blobs of dams and – if the bird hasn't spotted them elsewhere already – the smaller dark blobs of our Angus cattle.

In a matter of minutes, the bird will have surveyed an area smaller than Luxembourg and even Gibraltar, but bigger than Monaco and Vatican City combined. For many years, its population: my parents. And now, for two days every week, me too.

I began to equate Gollion with work. On Wednesdays and Thursdays, it was my workplace: I would arrive at 8 a.m., perform assigned labour with my father, then leave. (There were breaks for morning tea, lunch and afternoon tea, and Mum would occasionally invite me to stay for dinner, but Dad's expectation was that I would return to the city at sundown and resume my own life.)

But Gollion was also the *object* of our work. Everything we did was contributing to a greater whole. The farm was the sum total of Dad's projects, and had been since he first sank a shovel into its crap soil. Until now, I had walked Gollion's hills and taken what I'd come across as a given. The placement of a copse of trees here, a patch of lush grass there: they were incidental scenery, part of the amorphous 'environment'. I came to understand that much of what

I thought was 'natural' was in fact manipulated; incidental features that made the countryside 'picturesque' had been carefully designed. It was a grand conceit.

A poem about this was written by Alexander Pope, the eighteenth-century poet who as a child survived being trampled by a cow and went on to become a bard of the English countryside. In *Epistle to Burlington* (addressed to Richard Boyle, third earl of Burlington), Pope advised the aristocrat, an early practitioner of landscape architecture, to 'Consult the Genius of the Place in all'. Where in a garden does the rain settle? What is the aspect? Which parts receive morning sun, and which afternoon? For Pope, harnessing the *genius loci* – the spirit of a place – unleashes a creative force that 'Paints as you plant, and, as you work, designs.'

What was Gollion if not a large, landscaped garden? A microstate-sized park?

'You want to plant your trees on your ridges,' Dad told me one afternoon, overlooking a hilltop plantation of mature eucalypts that I hadn't realised was a plantation. 'When rain hits a bare ridge, it runs downhill without soaking into the surface; a timbered ridge helps trap moisture, so it slowly soaks in. Ridges are your recharge areas.'

In a notebook, I wrote: 'Ridges = recharge areas.'

Similarly, Dad told me that in this undulating landscape, he'd designed his paddocks to run up and down hills, not across them. This was to force livestock to exercise as they grazed, and to make them return nutrients (in the form of grass) from the bottom of ravines, where rain naturally washes them, to the tops of hills (as dung). Sheep especially, said Dad, like to make camp on hilltops, to catch the morning sun. In this way, pasture would have a chance to grow everywhere on the property – the bits not covered with rocks, at least. 'You want to aim for 100 per cent green grass 100 per cent of the time,' he said. What you didn't want, he said, was ground with no grass, which radiates heat, doesn't fatten cattle and deflects rainwater. 'Bare ground is your enemy.'

To the notebook, I added: 'Bare ground = the enemy.'

Even what I'd taken for rubbish I now saw had been put to strategic use: shredded documents and old cardboard boxes were spread across pastures to feed microorganisms, and branches of pruned trees were dragged onto patches of bare ground to create a humid microclimate so other plants could take hold. (Of course, there was rubbish that served no use, too: Gollion had *eleven* cars in various states of roadworthiness, as well as the hulks of antiquated tractor equipment. It wasn't as bad as some Australian farms, which seem to sprout nothing from their soil but decommissioned whitegoods.)

We did nothing but fix fences for weeks, because that was the job Dad deemed hardest to do with one hand. When I'd tell city friends I'd been 'fencing' again, they'd invariably think I meant swordplay – one even asked: 'Foil or *épée*?'

I grew used to the donger (the trick was to release it the moment before impact on the steelie to avoid its jangling vibrations). I was even allowed to use the mechanical driver. The knots used to tie wire to corner posts and to itself took more time to master. No matter how many verbal learning aids Dad repeated ('In, out, under, over'; 'the monkey swings round the palm tree ...'), I've always found these hacks unhelpful. But by copying Dad exactly, standing behind him and holding the wire as he held it, I got the hang of these, too. One evening back in Canberra, after working on the farm, I emptied my pockets and found two bits of loose wire. I executed a perfect figure-eight knot to see if I could, and put it on my bedside table, where it remained for years.

The same knots are used to link and tie off barbed wire. Whereas plain wire was designed to construct fences, barbed wire was designed to hurt humans and is remarkably successful at this: after a few seconds of handling it, even with pliers, my knuckles would be nicked red and beading with blood. Papa, who my father says used to spend most of his time fixing fences, had a series of scars covering his legs; it was only years after his death that I learned they were made by barbed wire – not from one of Papa's fences, but from soldiering in the jungles of Borneo during World War II.

Papa's name was evoked surprisingly often while we worked. One of the last steps in building a fence is to strain the netting, the prefabricated wire roll that comprises the bottom half of the fence. It consists of rectangles that are too small for livestock to pass through but large enough for birds and small mammals. To do this we attached a homemade wooden clamp to one end of the netting, then strained the clamp, via two wires, to the towbar of the ute, which was parked nearby. The taut net was then pinned to the corner post with giant staples and its six strands of wire cut, before each one was tied off around the corner post. This was known as 'Papa's trick' – he had designed the clamp, the prototype for our own.

It was clear that Papa – my father's late father-in-law – had been an important source of knowledge as Dad started out as a farmer. This realisation that Dad hadn't always been an expert made him less imposing to me. (Although, with his bald head, aquiline nose and plastered hand tucked into his jumper, he did resemble Napoleon inspecting his troops when he looked over my work in the paddock.)

There was something else Dad evoked surprisingly often: the concept of *regeneration*. This was what happened when you allowed the genius of the place to reach its potential: wooded ridges that worked as *recharge* areas; denuded hillsides that, through the clever placement of fences, *returned* their lost nutrients.

It made me re-evaluate what Gollion was for. A place for Mum and Dad to live and raise their kids, obviously. And a place to keep Dad occupied with projects. But this was also a business: sunlight and rain were turned into food, a commodity that could be sold at a profit. To improve this business, they had a choice: import external inputs (hay, fertiliser) or make their existing inputs work more efficiently.

But managing a landscape also came with a moral responsibility. 'I'm the custodian of this patch of the Earth for now,' Dad said, 'so I've got to look after it while I'm here.'

I'd always assumed Dad subscribed to the concept of sustainability, but now he described it as 'old thinking'. It denoted 'putting the brakes' on antiquated, harmful practices – clearing the land of trees, spraying chemicals – without actively reversing the damage. A farm whose ecology has been denuded through historic, discontinued practices, and that now sustains a steady (and lower) agricultural output, is still a denuded farm. 'Treading water' was the analogy my father used.

I have since come to think of sustainability in banking terms: the taking of an annual dividend (a harvest) without overdrawing on the natural capital required to produce it (the soil, the water, the air, the wider ecology). Conventional agriculture has historically prioritised the former over the latter. When nutrients leave a farm without being returned – in the meat of beasts, in the flesh of fruit, in the grain of crops – a deficit occurs. Left unchecked, harvest upon harvest, a deficit becomes an overdraft. Dividends shrink and the farm becomes bankrupt – it is unsustainable. But if sustainability is just the ability to balance the ecological books (by fertilising, irrigating, using 'cover crops' in fallow fields), that's a low benchmark. Natural capital isn't being built; it just isn't being reduced.

Regenerative agriculture aims to build the natural capital of a farm *and* take a dividend. It works from the premise that nature is complex, self-organising and self-repairing – but that in most of the world's ecosystems, it has been so badly damaged that human intervention is needed.

Buying worm castings and spreading them on your pumpkin patch: sustainable.

Creating a pumpkin patch that attracts worms so that they will move in, spread their castings and multiply their numbers, so that *more* worms leave *more* castings: regenerative.

Regenerative agriculture, Dad told me, was what we did at Gollion. In his words, it let nature 'work for you'. The goal wasn't only to grow food and make money, but to revive the landscape as you went.

The question was, revive it to what?

# 2

'The farm' was really three farms. Since 2000, when Gollion expanded for the second time to its current 650 acres (in 1994 it grew from 200 to 450), my father had gradually unified each part, fencing and re-fencing pastures until every paddock was roughly the same size, then grazing his cattle in each according to a roster. (In the summer of 2003/04, when drought made the ground hard to dig, he fenced two kilometres in two weeks, giving himself tennis elbow and several bouts of what my mother suspected to be heatstroke.) But geographically, environmentally, legally (there were three title deeds) and *culturally*, each block had a distinct look, feel and personality. The blip of time they had been 'Gollion' failed to erase the marks of earlier custodians.

The original 200 acres bore the most obvious trace of my parents' toil, having been under their management the longest. The house they built in 1983 was a low-slung brick bungalow with a tin roof (so my father could listen to rain fall on it in bed) and a pine interior that glowed orange in the light cast by the fireplace (Vincent family home videos have something of the *Twin Peaks* look about them). The whole thing was delivered as a kit on the back of a truck and assembled on a bulldozer-levelled clearing cut into the side of a hill; characteristically – 'I'm a big picture kind of guy' – Dad didn't read the instructions closely and put the house together back-to-front, the large windows designed for north-facing winter sunlight placed on the south, plunging the building into darkness for much of the year.

What it lacked in light the house made up for in shelter. Tucked into its hillside out of the wind, the remaining flattened ground was covered with a cottage garden, beds of hydrangeas, camellias, wisteria and a brick path, Dad's uneven paving resulting in the unintended charm of a cobblestoned Old Town. Up a flight of steps and back onto the hillside, the 'lawn' (an arbitrary distinction: pasture grazed by mouths *and* the mower) was ringed by seedlings of ash, birch, elder and maple; a glasshouse; a berry patch; rows of vegetables.

Between the veggie patch and the chook shed, a gentle slope became the first orchard, its eastern aspect designed to catch the morning sun (Dad got that one right). My parents have said they wanted to be 'self-sufficient' in fruit, but how many children were they planning to have? The orchard was planted with a variety of seventy trees to gradually bear throughout spring, summer and autumn; from late November, when the cherries started, to early May (quinces, apples, pears and hazelnuts), forays to the supermarket produce aisle stopped, resuming in late winter only when the pantry's bottles of home-preserved fruit (peaches, plums, apricots and cherries, often with their tooth-cracking pits accidentally included) had been exhausted.

The orchard was protected from birds and bats by a giant net, sewn together and suspended twelve feet in the air by wires that ran between metal posts concreted into the ground. The design was inspired by a visit Dad had made in the 1970s to an orchard established during World War II by Italian soldiers detained as POWs and put to work as unpaid rural labourers.

A picket gate on the western side of the house marked the end of the garden and the start of the 'back paddock'. It was wilder there, exciting and faintly menacing. My sisters and I were allowed to play in the back paddock (then most of the 200 acres), but we understood our safety wasn't guaranteed. The dangers were real (brown snakes were common) and imagined (late one night in the early 1990s, when a convicted murderer was on the loose having escaped from Goulburn jail sixty-five kilometres to the north-east, my mother commanded my father – still in his pyjamas – to arm himself with the .22 when she heard branches snapping by the back gate. Of course, it was a cow).

In my earliest memories of the back paddock it is always summer, dry grass the tawny brown of a lioness. The trees – stringybark, yellow box, scribbly gum – formed an open woodland, their trunks burnished with the lanolin of bygone sheep. This was an unmistakably Australian landscape: trees that dropped their bark instead of their leaves and were scrappy and disordered compared to the neat lines of tagasaste plantations that crisscrossed the slopes nearer the house. Just as I suspect Dad was subconsciously channelling the storybook hedgerows of the Home Counties and Mum the flower gardens of Beatrix Potter stories, the trees we drew at school were stylised lollipops – more like oaks than the eucalypts we saw every day. Even in the 1990s our view of what a countryside *should* resemble was English. If anything, the back paddock and the open plains beyond it resembled the prairie of *Dances with Wolves*, which was on high VHS rotation at our house in those years. It seems obvious now, but even Kevin Costner serving coffee to a party of Sioux braves in his log cabin couldn't prompt the self-reflection required to understand that our ambivalence towards the back paddock – wilder than the rest of the farm, and somehow quieter – lay in the awareness we felt there of the dispossession of Aboriginal people. That reckoning wouldn't come for decades yet.

The hills of the back paddock rolled westwards out of a floodplain, each one higher than the last. Too steep for cropping and long ago leached of most of their topsoil, they were of marginal agricultural value. But here too there were archaeological reminders that my sisters and I were not the first white settlers to walk these hills, stomping and clapping as we moved to scare away snakes.

The gully that split the back paddock in two was clogged with a rusted car chassis, tipped sidewards and spewing out its seat springs. There was an old well, which we filled up with rocks, and a windmill just outside our northern boundary that screeched in the wind, scaring me at night.

With distance from the house also came freedom: you could be naughty in the back paddock. I spent much of my childhood there making 'bases', crude lean-tos of logs and bark from which I launched covert slingshot assaults on kangaroos and kookaburras. (I never hit any.) Later, in the twilight of my boyhood, I kept a cache of Cindy Crawford lingerie ads there, ripped from glossy magazines stolen from my sisters' rooms. I liked to look at them out there in the bush, without knowing why.

Arson was my sister Sarah's preferred expression of rural delinquency. She would light tussocks under the flimsy pretext of clearing them for pasture, then watch them crackle and burn with satisfaction. One summer she set fire to an entire hillside; Dad, a former volunteer firefighter, managed to douse the conflagration with a water tank he kept on his ute before the actual fire brigade was required.

From the top of the back paddock our house glared far below, a small rectangle of silvery tin not yet faded by the sun. The Yass Valley spread out beyond it, a floor of swaying grass rarely interrupted by farmhouses and poplar trees, then turning blue with eucalyptus haze as the other side of the valley ended in slopes of bushland. Giant powerlines, the kind with V-shaped chests and puny T-rex arms holding up the wires, blighted the landscape as they marched across it. (Dad: 'I don't even notice them anymore'; Mum: 'We talked about it when we bought this place but decided in the end that nowhere is perfect'.) They buzzed when it rained, leading me to believe that if I strained hard enough, I could eavesdrop on phone conversations passing along them.

Looking the other way, to the west, the back paddock seemed to go on forever, but in 2000, seventeen years after they arrived, my parents bought the 200 acres beyond it, extending the range of my peregrinations even further. The land out there (we called the new boundary paddock Rear End) felt different, cleared of trees but sown with fewer European grasses – which were known as 'improved' pastures by previous settlers, who saw little value in native plants, even as animal fodder. The grasses beyond the back paddock were overwhelmingly natives – they remained unimproved.

There were the ruined foundations of a nineteenth century shearing shed and a thicket of tree-of-heaven where a garden had once been. The homesteaders had chosen the site well: it sat in a kind of natural amphitheatre that directed rainfall to the home paddock and was a ten-minute walk to collect freshwater from the Murrumbateman Creek.

By the time my parents bought this land, clear-felling and overgrazing had turned the creek into an ephemeral stormwater drain: when rain fell it surged through and had disappeared completely within hours. You couldn't ride a motorbike through the paddocks there in spring for fear of hitting tree stumps, handsaw marks still visible like the crinkles on pre-cut frozen carrots.

The eastern third of Gollion, on another floodplain closer to the village of Gundaroo, was different again. By the time my family moved next door, Fernleigh farm had been reduced to a quartet of blazed paddocks (nobody had thought to plant a tree), a tumbledown homestead and the various sheds that had once supported a much larger enterprise and the workers to sustain it.

Gazing down the hill from our house was a glimpse across the Corrugated Iron Curtain. My family acquired Fernleigh in 1994 from a family still living in the nineteenth century. Mr and Mrs Whyte were barely older than my parents but in my mother's understated assessment were 'old fashioned'. They used an outdoor dunny (an endless source of childhood fascination: *What about redbacks? What happens if you need to go in the middle of the night?*), did their laundry in a wash copper and ridiculed their daughter for *finishing* high school; these were people who milked a cow daily on one of those three-legged stools and threw a pail of slop to a pig each morning to fatten it up for Christmas.

Mr Whyte was a moon-faced man with a head even balder than my father's, but whereas Dad opted for the koala-ears style of unruly tufts either side of his shiny dome, our neighbour's pate was dappled by an unctuous combover. He was the council dogcatcher, a position I misunderstood to mean he would seize *any* dog with the king-size butterfly net he must surely use, so I avoided him for fear of losing a pet. Mrs Whyte I have no memory of, but family anecdote records that soon after we moved into the district, she rang my mother and passed on a message: 'My husband says that if you don't get your ducks off our dam he'll shoot them.' They hosted an annual bush dance in their shearing shed each Christmas; my sisters and I would pedal our bikes to the windbreak of pines that separated our farm from theirs and watch the dust rise from the cracks in the roof as the wooden floors beneath were tapped and stomped.

When they moved out, the Whytes left us everything that wouldn't fit in the back of their ute. The place came fully furnished for the needs of frontier life: it had a kitchen table *and* a horse-drawn cart; a garage *and* a blacksmith's forge. Sorting the trash from the treasure was like coming upon one of those open-air heritage museums after the staff have taken off their bonnets and driven home. Suddenly my make-believe jaunts in the back paddock were accessorised with a rusted sickle and a pair of hand shears.

Apart from the iron roofs, the sheds and shacks had been built from materials found or forged on site: you can still see the scars in yellow box trees where beams and bark had been adzed, then rolled and dragged cross-country to be erected once more in a different form.

By the time I discovered them, most of the buildings were in a state of disrepair. The blacksmith's hut was on its way to the sleeping position it now occupies, its dirt floor covered with rusted kerosene tins that once served as spark-proof cladding. Towers of hay bales supported other buildings in lieu of beams, and one day I fell through the rotten floorboards of the wool shed. In one worker's shack, long ago taken over by swallows, I ran my hands over the dusty interior wall and discovered that someone had prettied it up with a newspaper – from 1896.

The Whytes even threw their ten horses into the deal. It would have been a hassle to relocate them, they explained afterwards, so they mustered each beast into an erosion gully then shot each one in turn. That was just how things were done around here.

In 'The Gundaroo Bullock', a 1917 poem by Banjo Paterson, this part of Australia is a cruel, desperate place, home to unprofitable farms and indebted farmers:

> Far away by Grabben Gullen, where the Murrumbidgee flows,
> There's a block of broken country-side where no one ever goes;
> For the banks have gripped the squatters, and the free selectors too,
> And their stock are always stolen by the men of Gundaroo.

Paterson wrote 'The Gundaroo Bullock' after 'Waltzing Matilda', and its plot rehashes the more famous bush ballad: a landowner's animal (this time a bullock instead of a jumbuck) is stolen by a hungry outlaw (Morgan Donahoo, 'the greatest cattle-stealer in the whole of Gundaroo'). The troopers are called (the law always collaborates with the establishment in Paterson verse), and the suspect is pulled out of bed and asked to explain himself. ('Now, show to us the carcass of that bullock that you slew – The hairy-whiskered bullock that you killed in Gundaroo.')

But all the police find in Donahoo's harness cask are bits of unidentified fur and sinew. He comes clean:

> The times are something awful, as you can plainly see,
> The banks have broke the squatters, and they've broke
> the likes of me;
> We can't afford a bullock – such expense would never do –
> So an old man bear for breakfast is a treat in Gundaroo.

From the 1890s, when Australia was in the grip of its worst drought since European settlement, 'Gundaroo bullock' became widespread shorthand for koala meat in Australian English. Native foods didn't have the cachet of sustainable consumption they do today; Morgan Donahoo wasn't eating koala to reduce his food miles.

The Gundaroo district was settled in the early nineteenth century when waves of oxen drays fanned out across New South Wales from their beachhead at Sydney. In early December 1820, explorers Joseph Wild and Charles Throsby Smith came upon a place the local Ngunnawal people told them was called 'Candariro'.

Like in much of colonial Australia, the land-grab that followed was initially unregulated, with well-connected 'squatters' occupying as much Crown land as they could and the scraps left for the likes of Morgan Donahoo. (Unsurprisingly, the koala, now an endangered species in New South Wales, is locally extinct. The last 'Gundaroo bullock' seen on Gollion was a dead baby, retrieved from the bottom of a tree by my Uncle Simon's Rhodesian ridgeback–pitbull cross, Sybil, in 1994, and eaten with gusto on our back lawn. It took us a moment to work out what it was, it having quickly lost any duty-free cuddliness in Sybil's wet maw.)

In time, the freeholders who followed in the squatters' wake became known as 'cockies', unable to show much for their small acreages except the screeching cockatoos who tore out the green shoots of crops in their fields and roosted in the remnant stands of gum trees above.

Mr Whyte's great-grandparents were cockies, and the first white people to turn pink farming what is today Gollion. In 1849, George Whyte, a free settler from Somerset, arrived in Sydney and found work as a farmhand outside Gundaroo, where he also found a wife, Mary.

George's brother Hugh White sailed to Australia in 1856 and settled nearby. When he arrived, he noticed that his big brother, who was illiterate, had had his surname misspelt by the colonial authorities – a mistake that would allow one family to claim they were two, bypassing laws on the concentrated accumulation of farmland.

George and Mary built a homestead of slab walls and stringybark roofs and busied themselves ringbarking trees and digging wells (including the one we filled up with rocks). Their principal crops were wheat, wool and children. ('While the brothers White and Whyte may not have contributed to literacy in the region,' a *Canberra Times* article from 1979 quipped, 'they helped in other ways. According to Canberra's City Manager, Mr Cecil Thompson, after six generations in Australia they have more than 2000 descendants, Mr Thompson among them.')

With enough children to fill a school, George and Mary successfully lobbied the New South Wales government for one; it was built in 1874 near their house. The building was called a 'disgrace' by a government inspector seven years later, and a new one was built of stones two kilometres to the west, at what is now Gollion's letterbox.

That school burned down in 1915, and its blackened stones were purchased by Mary and George's son, Joseph, to build his house across the road at Fernleigh. Joseph was the dogcatcher's grandfather.

By the 1980s and 1990s, when my sisters and then I started primary school, the nearest school was ten kilometres away in the village of Sutton. We'd wait for the bus to the new school beside the old school, kicking up inkwells and bits of writing slates with our heels.

Mrs O'Toole, the driver of the minibus dispatched to collect us, was a permed alpaca breeder with three pink tracksuits to her wardrobe. She scowled at us through her crow's feet in the rear-view mirror, always alert for backseat miscreants 'being smart' or 'playing up'. Her favourite phrases were 'STOP THOSE CROCODILE TEARS!' and 'THAT'S ENOUGH OUTTA YOU!' She ran down snakes sunning themselves on the bitumen and reversed back over their broken bodies just to be sure. Sometimes she'd stop the bus completely and jump out if her target made it off the road, and we'd all shimmy to the verge-side of the bus, causing the vehicle to lean towards Mrs O'Toole, a blur of pumping polyester, as we watched her selecting, weighing up, and then lobbing rocks into the grass. When she eventually dropped us at school, yanking the lever to jolt open the pneumatic doors, the honour roll on the gates said it all: Whytes and Whites; Whytes and Whites.

This was the world my parents entered in 1983. A community who hadn't moved from the valley their ancestors arrived in the previous century, and who had barely moved on from the lives of those ancestors. A community *grounded* in place. In its kindest definition, the adjective conveys one who is 'sensible' and 'down-to-earth'; having one's 'feet on the ground', connected to country. But to be grounded, as bored teenagers know too well, is also to be stuck, imprisoned, locked in your bedroom while outside, life happens without you. To be grounded is to be somewhere, not the everywhere you'd prefer to be.

The decision to build Australia's national capital of Canberra nearby in 1913 distorted this demographic. Suddenly a new wave of settlers arrived, a different kind of nation-builder. These people were educated urbanites from Melbourne and Sydney: public servants, bureaucrats and boffins, bowler hats and later lanyards flying off and about them on the windy streets of the empty city. My father among them, these people were university graduates, and the opposite of grounded: highly transient and socially mobile, ambitious careerists who had left the bright lights to do their time in the nascent capital because it would allow their careers to 'take off'.

Like Mr and Mrs Whyte, our neighbours in those days were mostly the age of my parents, with kids the same age as us. But it seems wrong to label them baby boomers – a cultural demographic as much as a generational one. By the 1980s they didn't have enough land to just farm, and supplemented their incomes with off-farm jobs as tradies and truckdrivers, publicans and shearers.

Opposite Fernleigh: another Whyte landholding. The weatherboard homestead of Westmead Park, with its wraparound veranda, has since been restored by cashed-up tree-changers, but I still associate the thwack of its screen door with the previous owners. Jack Whyte, George's grandson, was an old widower by then.

Jack died before I was ten ('His liver was buggered,' Dad told me recently; 'He lived off whisky and milk arrowroot biscuits,' added Mum) and was succeeded by his son Dusty, a woodturner with a gravelly voice and big, twitching moustache. Any 'cheek' out of his kids and Dusty would undo his belt and give them a thrashing. One night he pulled a gun on his wife and fired it above her head. The wife moved out. But Dusty considered himself a good Christian: he towed our school bus across the creek in times of flood and kept a picture of Jesus on the kitchen wall.

The children on our bus run who weren't Whytes largely lived on White Road. Paved and closer to Canberra, this was the front line of a kind of rural gentrification, with the original families subdividing and cashing in, making way for more people like us. Actual peasants were replaced by lifestyle peasants; people who had hobby farms of olives and grapes; people who fattened pigs for the cred as much as for Christmas.

But the biggest change to the landscape in my lifetime has occurred to Gollion's south, where the outer suburbs of Canberra have steadily advanced. By the early 2010s they were abutting the border with New South Wales, not three kilometres from Gollion's boundary.

In the Ngunnawal language, the word *gungahlin* either means 'white man's house' or 'pile of rocks', and in its modern, capitalised usage, both are applicable. In its latest incarnation, Gungahlin is the second-fastest growing part of Australia: eighteen suburbs built atop the old sheep station; tens of thousands of White Man's Houses occupied by a McMansion with a rumpus room and a columned portico: that most succinct expression of assimilation to the Australian middle class.

The border between Gungahlin and the bush is sudden and severe: the streetlights end and the darkness begins. There is even a kind of border wall separating the last house in the suburb from the first stand of woodland. Gungahlin locals only venture into the bush to abandon stolen vehicles or dump more prosaic refuse (mattresses, grass clippings, even a rowing machine).

Between the commuters and the dumpers, it's getting pretty busy. What were dirt roads when I was a kid are now largely sealed. Only the odd stringybark has survived the march of Canberra's northern fringe ('The perfect place to nest', the realty billboards proclaim).

In 2005 or thereabouts, I made the conscious decision no longer to greet every driver between Gollion and Gungahlin with what I have since learned is called the 'phatic finger', the (barely) raised index finger above the steering wheel one driver gives to the next as furtive acknowledgement of existence. The traffic was getting too heavy. You can't be going handing out waves willy-nilly every two minutes. It'd cheapen the effect.

I once asked my father what he thought was the most important skill of farming. He replied without hesitation: 'Paying attention'. He didn't mean it in the sense of 'paying attention in class', something he urged me and my sisters to do while we were at school, lest we 'end up fixing potholes' (or being a farmer), but attention in the sense of *engagement, noticing, reading* the land.

An obvious example of what he meant occurred one morning when I accompanied him to the Fernleigh end of the farm. It had rained heavily the night before, and the paddock where the cattle were was, in the words of rural road-signs, subject to flooding. Cattle will seek out higher ground in a downpour, he told me, even if that means breaking down fences. When we arrived, one calf was hanging upside down in the fence, its left back leg stuck where the top plain wire was stapled to the netting. The calf's eyes were rolled back in its head. It was alive, just. We cut it out of the fence with pliers and propped it upright. It kicked me in the shin – hard – then galloped away.

Dad said, 'If we'd have arrived a couple of hours later, that calf would be dead now.'

The paddock we were in was parallel, across Westmead Lane, to Westmead Park. As the calf joined the rest of the mob, Dad gestured with his head across the road to the white homestead with its dark pines.

'Jack Whyte was an odd man, but a good farmer – an

old-fashioned professional farmer. He'd check his whole farm every morning, then knock off by 11 a.m. – and hit the whisky.'

I asked him what he meant by 'old-fashioned' and he replied 'observant' – Whyte senior monitored his property with his eyes and reacted with his hands and his mind, not with the 'latest and greatest gear'. Jack's son Dusty, Dad told me, 'blew it' once the old man died; 'you'd always see Dusty on the front of *The Land*, posing with the most expensive ram at the sale. He didn't understand that $100,000 tractors don't make you money ...' The family holdings frittered away by the spendthrift son: a cautionary tale of farm succession.

But the attention my own father paid to his farm went beyond checking the livestock each morning. I'd never noticed before, because I'd never walked the farm with him, how he often crouched and ran his hands through the grass, as if he was looking for something he'd dropped.

'Tell me of what plant-birthday a man takes notice,' wrote American naturalist Aldo Leopold, 'and I shall tell you a good deal about his vocation, his hobbies, his hay fever, and the general level of his ecological education.'

My father was a primary producer, and as he ran his hands through the grass he was taking notice of the most palatable – and *productive* – grasses for his herd: the clumps of phalaris he had sown by hand; chicory sown with the tractor; fat-leafed paspalum that is a summer-active grass, contributing to his goal of making Gollion '100 per cent green, 100 per cent of the time.'

But he noticed other 'plant-birthdays' and taught me to question why something grew when and where it did. Thistles weren't 'weeds' to mindlessly dig out but a 'sign of previous poor land management': they thrived on bare ground, including disturbed soil that had been overgrazed, cleared or tilled; farming this way was 'pushing the envelope' – it was unsustainable. A thistle's taproots burrowed deeply for nutrients, meaning they could grow in soils that plants with shallower roots could not – their presence showed that the topsoil was deficient. Cape weed told a similar story, as did Paterson's curse, the purple flower that is toxic to horses but can provide emergency fodder to ruminants, whose four stomachs help break down its alkaloids ('Remember the early settlers called it Salvation Jane').

In this way he introduced me to the concept of plant succession, whereby denuded land is first colonised with fast-growing annuals that cover the bare ground and bring up nutrients to replenish depleted soil before dying and becoming mulch, thereby creating more suitable conditions for something higher up the chain of succession to take root, a perennial grass, say, or one that can't set root in bare ground. The process gets more ecologically complex each step of the way, from lichen to trees. Viewed this way, 'weeds' weren't something to destroy but to understand. They were doing a job, scabs on a scarred land, and would go away naturally if you managed them correctly. If you picked at the scars (more overgrazing, more land clearing, more ploughing), you would get more weeds.

The 'job' was regeneration: helping the land to become biodiverse, covered in photosynthesising plants growing on healthy soils that stored water. That would make the farm more resilient to drought, more healthy and pleasant for its human and non-human inhabitants, and a more effective carbon sink to combat climate change. In Dad's words, 'We aren't trying to build Jurassic Park here': the massive changes wrought to the Australian environment since the arrival of Europeans made it impossible to bring back all that was lost. What we could do was revive the function of the landscape, if not its exact form. While the earlier generations of farmers here saw themselves as taming an unruly land, we, says Dad, 'play on nature's team'. 'Our job is to see what we can do for nature, then get out of its way'.

Kentucky farmer and writer Wendell Berry calls land management a 'conversation with nature', beginning with three questions: 'What was here? What will nature require of us here? And what will nature help us do here?'

What was here in 1820 when explorers Joseph Wild and Charles Throsby passed through? The answer could be gleaned from contemporary landscape paintings, perhaps, or from the explorers' own journals, or from the oral histories passed down from the Indigenous people they met. But it could also be learned reading the same landscape two centuries on: cockatoos still roosted in Fernleigh's gum trees, especially the hollowed-out victims of the early settlers' ringbarking. The filled-in well now sits in a dry ditch, but it would've once been the location of a spring, and was, Jack Whyte told Dad, the only well in the district not to run dry during the fin de siècle Federation Drought. For government weed inspectors, a blackberry bush was a problem; for Dad, the one growing on Gollion's eastern boundary was a 'rain gauge'. Blackberries only grow in areas that receive 700 millimetres of annual rainfall or where watercourses kept the ground wet; given Gollion's rainfall is around 630 millimetres, the bush was a sign of a once-functioning creek.

One farm day, I was on my way out to Gollion when I stopped for a coffee. The barista asked what I had planned for the day. I told him I was heading to my family's farm near Gundaroo to help my dad do some cattle work. Beef or dairy cattle? he asked. Months earlier I wouldn't have judged the question – I may have asked it myself. But as I walked away, I found myself thinking it odd that someone who steams milk for a living didn't know the Southern Tablelands of New South Wales were too dry for dairies, which require year-round green grass.

# 3

Like many Australian farms, Gollion observed a gendered division of labour. Four-legged animals, trees (but not shrubs), lawnmowing (but not flower picking); mechanics, construction; keeping the fires burning and the taps running: this was Dad's domain. Mum's place was in the kitchen, at the clothesline, kneeling in the garden and caring for the chickens. (The hygiene of their water troughs, technically plumbing, was Dad's job. Collecting their eggs and throwing them scraps: women's work.)

From a young age I noticed the transformation that occurred when Dad crossed the threshold and entered the house. Kicking off his boots on the back veranda and sliding the glass door open with a clatter, my father – the most multi-practical, useful person I have ever met – would revert to a state of newborn dependence.

He rarely cleaned, and if Mum asked him to, he would ask her where the cleaning products were kept; he rarely cooked, and when he did (a set menu: bacon and egg pie or pea and ham soup), he fussed over which utensils to use and, sometimes, what they were for. (Once, when I was in high school, Mum was enjoying a night off from cooking when he stormed into the living room, boiling water sieving through the stainless-steel steamer he held in his hands, and yelled, 'Who drilled HOLES in this SAUCEPAN!?')

The one 'domestic' duty he accepted without complaint was the carving of roast meat. I had always been confused by this exception – there seemed no great secret to carving, and if there was, Dad's misshapen hunks suggested he didn't possess it – but one week in early spring, it occurred to me why.

It was time, my father ordained, I learned to butcher a sheep. The orchards' first blossoms – almond and cherry – had burst, but the nights were still cold, allowing a carcass to hang without risk of spoiling.

Gollion's flock of two dozen sheep were dorpers, a South African breed from the Karoo plateau north of Cape Town, which has an arid climate that makes them suited to inland Australia. They don't require shearing (their winter fleece is shed each spring) and are more like goats in their wide-ranging appetite. At Gollion, they work as lawnmowers in the paddocks around the house, snuff out windfallen fruit in the orchards after harvest, then fertilise each tree with dung.

They were my parents' main source of meat.

I found Dad in the sheep yards below his tool shed, a faint sun burning off the last of the morning fog, where he handed me a bucket containing a handsaw and three knives – two thin and straight and one fat and curved like a scimitar – then disappeared back into the shed, re-emerging moments later with his .22 rifle.

I opened the gate into the race (a narrow, waist-high corridor walled with corrugated iron) and spotted what

Dad called the 'killer', a one-year-old wether (castrated male) he had selected earlier in the week because it was plump and healthy, marking the condemned beast's muzzle with a stroke of blue chalk like the spray-painted Xs on roadside trees destined for removal.

'Push 'em up,' Dad said.

I clapped my hands and the sheep ran into the race; I followed them inside and pushed them forward with my knees until they were bunched together at the end of the race. Dad side-saddled the wall and, with more a pop than a bang, shot the killer in the back of the head.

I thought: he had no idea. One moment, straining to keep his head above the butts of his friends like a swimming dog keeps its head above water; the next, lying in his own blood, twitching but very much dead.

Dad let out the rest of the sheep – some now bespattered, and having, to all appearances, no idea what had just happened – and I dragged the killer by his front legs into the yard. Dad handed me the scimitar and told me to 'bleed' the throat.

I got down on my knees and pressed the blade against the killer's throat, just below his Adam's apple.

'Here?'

'Yep.'

I pressed again, harder this time. No blood. Was the knife blunt?

'Keep going,' Dad said. He was crouching beside me. 'You need to cut through the windpipe.'

I knew when I had, because it gurgled like a burst water main, blood – crimson, almost black – pooling around us. The colour drained from the sheep's face.

We flopped him over so that he was belly-up. Dad picked up one of the straight knives and, with steady cuts, started skinning the wool around each leg. 'The trick is to keep tension on the leg you're working on,' he said. 'Jam it between your legs, and you can keep both your hands free.'

While Dad worked on the hind legs, I started on the front ones, copying the way he hunched over and tucked a hoof between his calves. I found that, once cut, the skin peeled back easily; the difficulty was in not also cutting the meat beneath. 'You're hacking at it,' Dad said. '*Cut* it, with fluid strokes.' Steam rose from the fat under the fleece, which, now exposed to the air, was turning from white to a creamy mustard.

When we had skinned the fleece back to the groin and armpits, I sawed off the head. The effect was immediate: without eyes on me (such doleful lashes!), I felt no judgement. The killer was killed: *he* had become *it*, the beginnings of a butcher shop's window dressing. We lifted it into the back of the ute and drove it down the hill to the meat hoist, the carcass and head jiggling with each bump, the grooves of the ute's tray running red.

On the grass by the hoist – a kind of seat-less seesaw – I cut the fleece back further to expose the chest; then, Dad indicating where with his forefinger, I sawed vertically down the length of the sternum, only as deep as the bone. With the tip of a knife, I made two holes in the tendons above the back hooves, poked an S-shaped metal hook between them, attached the hook to the lowered end of the hoist, and lifted the carcass vertically into the air. A chain on the low side of the hoist was clipped onto a post to support the carcass weight – between twenty and thirty kilos.

'Now punch off the rest of the fleece,' Dad said.

I made paddles of my hands and slid them along the sheep's flanks, between the inside of the fleece and the carcass. It was warm inside, and I felt fat catching under my fingernails as I forced my way through. My hands resurfaced either side of the neck. I entered again, this time punching with clenched fists until the fleece was loose enough to be yanked down and onto the ground. The naked carcass swung on its hook.

Dad told me how, when he was a boy, boxers used to work as butchers so they could train on the job.

'Now open up the chest without piercing the stomach.'

I pierced the stomach.

Here's what I was trying to do: while Dad used his forefinger to keep the stomach from escaping where I'd sawed through the sternum, I would extend the cut with a knife, upwards from the groin to the breast. But the stomach didn't want to stay inside – the part not immediately touching Dad's finger bulged out like an overinflated inner tube – and when pierced with the knife it let out a sigh. Liquid and half-digested grass splashed at my feet.

Dad said, 'That's why butchers use a metal ruler to push it in while they cut.' He was unfazed. 'Your Uncle Michael should be teaching you this bit – your grandparents made him spend a week with the local butcher when he was starting to go off the rails at school. From memory, I think that butcher ended up killing his wife ...'

Next, Dad pulled a piece of twine from his pocket and told me to tie it around the anus so its contents wouldn't spill on the meat when we removed the bowel. We yanked out the intestines, stomach, bladder and pancreas.

A running commentary was started on which offal was eaten by which nationalities, the 'delicacy' in question then tossed in the dirt by my father, the implication being that Australia was a rich country, so could afford to eat like one.

The lungs: 'If we were French, we'd eat these.'

The kidneys: 'Ever had a full English breakfast?'

The bladder, the trachea, the tail: 'In Papua New Guinea they'd eat it *all*: they'd clean it up and sell it on the street, two kina a bag – about fifty cents.'

This scene had figured in Dad's suburban childhood. It was once common for Australian families to keep a sheep in the yard of their quarter-acre block, and Dad told me he remembered visiting aunts and uncles in Melbourne who prepared their own lamb's fry. 'But then,' he added, as if by way of apology, 'that generation *did* live through the Depression.'

The bowel looked like a rice paper roll, little black balls of shit visible under a translucent membrane sausage. Dad flicked it as if it were a syringe. 'Looks like this one had a good last meal . . .'

We kept the heart and the liver, to be boiled for dogmeat. The rest was thrown in a ditch with the severed head and covered with a handful of builders lime to deter scavengers. The carcass was bundled into its body bag, a kind of giant pillow slip Mum had sewn up for this very purpose, then, back at the tool shed, hung from a rafter.

The following afternoon, we lifted it off its hook, transferred it to a wheelbarrow and onto Dad's butchery table in the back garden. I sawed it lengthways in half and, on Dad's direction, started cutting and cleaving cuts. There was a new running commentary: 'Your mum uses the shoulder for curries and stews'; 'Keep the leg intact and slow-roast it'; 'In fancy Sydney restaurants people pay top-dollar for a rack of lamb, but I've always preferred turning it into chops.'

Lambchops: they were tricky fuckers. It was a long way from the top of a cleaver's swing to the lamb rack on the table, and by the time it arrived, a gap had emerged between where I aimed the cleaver and where it landed. I tried steadying the rack with my left hand as I cleaved with my right, but this now meant I had a hand to avoid as well as a chop to cut. Dad pulled my left arm away: 'Better a crooked chop than a missing finger.'

He picked up a knife and started scraping the fat off the belly, something he said his mother used to request the neighbourhood butcher do when Dad was a boy, before she then minced the meat for Cornish pasties.

'Chuck us that piece of wood, will ya?'

'The cutting board?'

'Yeah.'

When he had finished scraping the flaps, Dad declared the job done. The table was covered with piles of meat, classified by cut, some more orthodox than others, a scene that gave sudden clarity to the expression to *butcher* a job.

'Usually I get your mother to do this next bit – I don't know where she keeps those little plastic Ziploc bags. Somewhere in the kitchen. She can bag and label the cuts and chuck them in the freezer.'

And there it was. Dressing a lamb became Mum's work when it became *lamb*. While it was a carcass, it was a masculine plaything; once *food*, it was feminised, at which point it could be taken inside the house and packaged, labelled and cooked by a woman.

Carving a roast presented a challenge to this dichotomy: it was done in a kitchen, but with the same knife and the same actions as on the butchery table. It was a passport between two worlds, that of outside (The World of Man) and inside (Here Be Dragons). The kitchen was a foreign place with foreign customs, where a few lengths of two-by-four, glued together, were passed off as something called a 'cutting board'.

My mother was a farmwife, but she was also a farmer. She was born in 1950 in a small town by the Murray River: a country girl with a *Boy's Own* childhood, she was helping her father and older brother muster sheep on horseback before she could ride a bicycle. 'You can get bridles where you have a little leading rein, and so Dad would be on one horse, and Mike next to him, and I would be being led on another horse. I was only two and a half, three when I did that!'

Papa had been a commando during the war and met my grandmother while at Officer Training School at Canungra, in the Gold Coast hinterland. 'We used to joke,' Mum recalled, 'that he'd march the men down to Brisbane to visit Ma – which was a very long march!' They married in 1946 and first lived at Craigie Mains, a sheep station in outback New South Wales and one of several properties owned by my great-grandfather.

For two years, Ma, a city girl from a bourgeois family (her mother, Norma Wilkinson, was in the first cohort of women to graduate from the University of Queensland, in 1914), lived in a corrugated-iron shack with no power, alleviating her boredom by learning to drive a car and making *mille-feuilles* – for when Papa tethered his horse to the veranda and tipped the sand out of his hat.

By the time my mother was born, the family was living at Aratula, another property of my great-grandfather's, in the Riverina. In 2002, ten years before he died, Papa wrote down his recollections: 'Aratula was a lovely property. Dark loam with a creek, Bullatale, meandering through the centre into most of the paddocks. The whole area covered with occasional large red gum trees, then gum forest between the lower paddocks and Murray River. It was a delight to ride a horse and work the stock in such a setting ...'

Papa managed Aratula for six years, Ma busying herself cooking, sewing and gardening, as well as looking after three young children, all the while challenged by life in the bush ('she'd go into the kitchen,' my mother said, 'and there'd be a tiger snake behind the stove or something').

In 1954, Aratula was sold and Papa again agreed to manage a family property. Koonje, in the Western District of Victoria, was nothing like Aratula: flat, bleak, wet and largely cleared of trees. At 1300 acres, it was exactly twice the size of Gollion, and many times more fertile. This is some of the best grazing country in Australia, and with beef prices rising and wool prices falling, Papa switched from sheep to cattle.

In Mum's telling, life at Koonje was more social for Ma (she played tennis and bridge with other farmwives and shopped in Melbourne and Warrnambool), and the climate, although 'very cold for a Queenslander', was more conducive to gardening.

But when Papa's own father died, Koonje was unexpectedly bequeathed equally to his six children. My grandparents went into debt to buy out Papa's siblings, something Ma resented because they were already independently wealthier than he was. (One sister had married a pastoralist whose landholding *was* in the ballpark of Belgium. 'It had its own *railway station*,' recalled Dad, ever the acre-counter.)

After my own parents were married, in 1972 (in Ma's garden at Koonje, Papa killing a snake shortly after the wedding vows), they lived first in Christmas Hills outside Melbourne (they named their block Aratula). Dad commuted to the city and Mum to a suburban girls school, where she taught agricultural science until my eldest sister, Eve, was born in 1977.

Like her mother before her, mine moved where her husband dictated, next to West Germany, where Dad worked for two years at a policy thinktank (the family now five, with my twin sisters, Lucy and Sarah, born in 1979); then, when Dad briefly found work in the public service, to Gollion via Canberra.

Koonje was sold in 1972, Papa breaking down and sobbing the day his eldest son, my uncle Michael, told him he didn't want to take it over. The more obvious candidates, my mother – keen on farming, a fresh agricultural science graduate – and my aunt Amanda (who spent her childhood in the saddle, helping Papa while her siblings were away at boarding school) told me they never proposed to their father that they succeed him at Koonje, and nor did he suggest it. Mum doesn't blame him: 'It wasn't the done thing back then.' At Gollion she would have the opportunity to manage a landscape.

I keep a photo of my mother from that time on my desk. She is squinting into the camera (my father has never been good at photography), her left eye barely a crack. Her chestnut hair, gleaming in the low sunlight, falls around her shoulders. It is winter at Gollion – yellow baubles of wattle flower, a blossom she describes as 'cheerful' – appear in the photo's bottom left. Behind her is a downward sloping paddock, ending at a windbreak of pine trees. The shot is remarkable for its absences: that empty paddock will soon become an orchard, a rose garden, a rhubarb patch. My mother looks happy and energised. There is work to be done.

Mum has never attached the prefix 'hobby' to this farm, because she understands the underlying insult. As well as the usual unpaid domestic labour, her influence is in every paddock: which trees were planted; the resulting birdlife there today. Long before me, she was the gate opener and cattle drafter. 'I missed that when you started working here,' she told me recently.

Those two worlds of inside and out were more fluid for her than for Dad. Perhaps because he is an economist by trade as well as a man, my father valued time spent producing commodities you *sell* over food that you *eat*, the result being a prioritising of cattle over kale, a new hayshed over a new henhouse. 'Your father seems to think that my time isn't as valuable as his': It was a familiar refrain of my childhood.

This tension was exacerbated now that I worked at Gollion. Mum felt that if Dad had a farmhand, so should she. I helped her prune the berry patch and mulch her roses; one day we built a new irrigation system for the shrubs by the gazebo. But when Dad would call for me from the tool shed or fire up the old ute and pause in the driveway, leaning over Suey Dog to open the passenger door for me, the expectation was that I'd leave her for him. Sometimes I lingered, but I always went.

It wasn't that Mum's needs were ignored by Dad – they were invisible to him. One day, he and I came into the house for morning tea. Mum was pulling a crumpet out of the toaster; as soon as she plonked it on a plate, Dad started buttering it and turned to me.

'Go you halves?'

Mum calmly left the room. Dad didn't realise she'd toasted the crumpet for herself until she slammed the door. But then, neither had I.

# 4

Farming in Australia largely involves swearing at the weather forecaster and swearing at the farm. There's the odd fence to fix and ute to bog, but these jobs can usually be completed in time for morning tea.

The first time I heard my father use the c-word he was holding a newborn calf in his left hand and a syringe of Ultravac® 5in1 (one jab protects against tetanus, malignant oedema, enterotoxaemia, black disease and blackleg) in his right.

'You stay still now . . .'

I would've been eight.

'Ya MOVED, ya CUNT OF A THING!'

Maybe ten.

It was an exciting new variation on a familiar theme: 'Mongrel bastard *of a thing*'; 'You bastard *of a thing*'; '(Little) fuckwit/head/er *of a thing*.' The common denominator wasn't the curse but its target: through an act that's not so much dehumanising as de-animalising, Dad would render the object of his frustration – a coochy-coo, six-week-old, Labrador-sized black Angus with sad eyes, lanky legs and velvety fur – a *thing*. Dad maintains it's a turn of phrase, but

I've never heard him turn it on a human.

I asked Dad recently if there was some Freudian clue to this construction – you can be as angry as you like with *property*, but abusing a *being* risks eliciting empathy. He admitted that farmers are at their angriest when working in the stockyards: marking, mulesing, branding, crutching and docking. When, in other words, they are hurting animals. His rationale was that animals panic in the stress of these environments, in turn frustrating farmers, who 'don't like wasting time'.

Time that could be spent swearing elsewhere on the farm.

I once met a dog named Fuckwit. Fuckwit's master, an itinerant agricultural labourer (albeit one sustained by servo pies and flavoured milk instead of damper and billy tea), roamed our valley doing odd jobs. He shall remain nameless, if only so I can state the obvious of someone who calls their dog that – he was one himself.

Naming a dog Fuckwit would've been a bit vulgar for my grazier grandfather. But he got into the spirit all the same. My aunt Amanda tells me that Papa used to have a tangible border on his farm, beyond which decorum ended and the gutter began: past the gate to the cattle yards, and only past it, you could say what you wanted.

My grandfather's idea of swearing was to call his cows 'old bitches'. My mother, an uncoordinated, dorky kid, would be called a 'clot of a girl' if she didn't open or close a gate fast enough. Even his favourite dog went by Biddy the Bitch after a telegram from her breeder arrived: 'Biddy the bitch arriving tomorrow's train.' Most offensive were the names of two of Biddy the Bitch's predecessors: Boozer and Darkie.

But the most memorable name for a working dog I've come across isn't Fuckwit, Boozer or Darkie. Maggot was a squirming grub of a pup. She transformed into a beautiful Fly.

We've never had working dogs on our farm. And our cows and calves didn't have names when I was growing up. But our bulls did. Cricketers were once in vogue. There was Gill'pie (in honour of the MCG scoreboard's inability to accommodate the full nine letters of Gillespie), McGrath (as angry as the paceman; not to be backed into a corner) and Warnie (when the cows were on heat, there was no better performer). There followed a period where our bulls didn't have names so much as character slurs (Young Fella with the Dicky Knee, Old Bloke with the Useless Prick).

Most maligned is the meteorologist. They're always called the same thing (liar) for the same reason (over-promising, under-delivering). For the farmers of this thirsty continent, there's nothing *fine* about a 'fine day'; a forecast of *showers* translates as 'a drop or two' and a *chance of showers* is generally a guarantee of more fine weather. 'It wouldn't hurt so much,' says Dad, 'if they didn't get my hopes up every time.'

Sometimes, farmers tip the farmyard confetti on each other. Dad likes to refer to contemporaries who've eaten a few too many CWA scones as having 'been in a good paddock' or 'gone to seed'. A farmer who succumbed to succession planning has been 'put out to pasture'. One dim-witted neighbour is said to have 'a few kangaroos loose in the top paddock', and not a cattle sale goes by without Dad referring to livestock agents – tongues wagging, responsive to whistles – as 'sheepdogs in riding boots'. To eat with your mouth open at my grandmother's table was to display 'shearer's manners'. Even the vocation itself is a source of slander: a crooked gate is said to be 'agricultural', while a shed that isn't level is still 'good enough for the bush'.

Then there's the infamous business card that is kept in a secret location on my best friend's family farm and only brought out on special occasions. Operating in the Bungendore region in the early 1980s, the Mob of Cunts were a team of fencing contractors. The slogan they chose to make a lasting impression on potential clients? 'We cum anywhere, anytime.'

Curiously, the one aspect of animal husbandry that seems to avoid dirty language is the one that, among many humans, generates the most. Bulls are said to 'serve' (or sometimes 'go over') cows and, if successfully 'joined', the cows are 'in calf'.

Keep this in mind next time you visit a farm. Nothing will mark you as an out-of-touch, inner-city coastal elite quite like asking if the bulls are currently rooting the heifers and, if so, how many they've knocked up.

It was the first week of September and I was bent over in the cattle yards, wondering if the calf who'd just shat on my hands would notice if I wiped them on its coat.

'You know that in Europe,' my father said cheerfully, peering over the post-and-rail fence, 'peasants used to sleep under the same roof as their livestock for warmth. All that dung and urine splashing around at the foot of your bed – it was a kind of central heating.' He was enjoying being out of the firing line for once, I think.

Gollion's 149 black Angus cows were served by three bulls the previous spring, and by the end of my first winter as a farmhand 100 calves had been born. Fifty-seven of those were males, which meant 114 hormonal timebombs were ticking between their legs and would need to be defused. By me.

Dad returned to 'drafting', separating the calves from their mothers. Aided by a series of holding pens and the odd whack from a length of poly-pipe, he released the cows into the adjoining laneway and directed their six-week-olds down a race to the crush, a kind of vice for livestock. There, securing four at a time, I leant in and vaccinated each one with a needle the way Dad had shown me ('grab them by the scruff of the neck and pretend you're a nurse') before 'marking' them with an electronic ear-tag and, after running an exploratory hand between hind legs, expanding a rubber castration ring around the bull calves' scrotums using a three-pronged pair of pliers, before contracting it once I was sure both testicles were below the ring. This last job takes the longest, a blind fumbling akin to the quest to find the plug in a share-house sink.

At six weeks old, a calf's gonads resemble table grapes, but a week after castration they would be more like sultanas. Starved of blood, they shrivel and stop producing testosterone, neutering a bull into a steer. This is done to prevent them serving their sisters, and because a steer is more docile than a bull. (Left unchecked, those grapes will grow to the size of grandfather clock weights. Strike them and you will hear how Big Ben chimes.)

When all the calves had been marked, they were reunited with their mothers in the laneway. Most rushed to bulging udders, their black muzzles soon white. Left in the yards were a few 'dry' (calf-less) cows, which would be run in a separate mob and sold in the autumn, some marked steers born last spring, and a cow in need of a new ear-tag.

Working the crush requires multitasking, like poaching eggs without burning the toast. When the cow with no ear-tag was in position, I slid the crush gate closed behind her with my right hand and, with my left, brought the jaws of the crush down over her neck. I next used both hands to work the two levers at the front of the crush, raising a bar under her chin to restrict her movement. She bellowed loudly and stomped. She poured out a shit.

I loaded the tag applicator – essentially a stapler for ear cartilage – and approached the cow front-on, but at the point of contact she jerked her head, causing the prongs of the applicator to misalign and the tag to be ruined.

I reloaded and tried again. Same result. Tag after tag was wasted; the cow was now frothing at the mouth and rocking the chin bar. A small heap of fudged tags lay in the grass, white on green.

On my next attempt, the cow knocked the tag applicator from my hand with her nose. I told her to *FUCK* OFF.

The go after that, another tag was botched by another late retreat. I asked her if she was FUCKING *KIDDING* ME*?*

I spat over my shoulder, reloaded the applicator and tried again, my left hand poised to grab the cow's right ear, my right hand moving the applicator into position.

And then she tried to bite me, so I slapped her in the face.

I was overcome with shame and let the cow out of the crush immediately, both ears defiantly unmarked. What was my excuse – that she'd been wasting my oh-so-valuable time? Thankfully Dad hadn't seen a thing. But I still wanted an answer.

What was the contract between the farmer and the farmed? I'd long known that 'cruelty-free' meat was marketing bullshit. From the moment they were yanked into the world with Dad's old wire strainers, through being marked, sultana'd, weaned and eventually put on a truck for the sale yards (or shot in the paddock if they 'broke down' first), being a denizen of Gollion hurt. And that's nothing to the horrors of factory farming.

'Animal husbandry': was that more spin? If one party dictates who lives and who dies, surely this is no happy marriage. By this definition, domestication means subjugation – if not bondage.

And yet. A decade ago, I spent a summer bobbing upon the Southern Ocean, living aboard and writing about a shipful of animal rights activists in pursuit of Japanese whalers. When I told one of them my parents had a cattle farm, he threatened to cut the fences and 'set them free'. What, I wondered, would that look like?

It was human will, the story usually goes, that transformed wild animals into domesticated livestock. Thousands of years of selective breeding has created species dependent on us. There are no wild herds of Wagyu, marbling their rumps as they roam the Eurasian steppe. Nobody sheared the feral merino ram 'Chris' while he lived in the Australian bush for six years. When he was found five kilometres from Gollion in 2015 and shorn, he had amassed a world-record fleece of 41.4 kilograms. (Shorn yearly, a merino's fleece is around five kilos.)

Domestication wasn't a political development, but an evolutionary one. Humans didn't impose it on animals so much as animals adopting it as a strategy for survival: domestication emerged when a few opportunistic species discovered, in the space of many generations, that their evolutionary longevity was better served through an alliance with humans than by going it alone. (This analysis is not mine, but the science writer Michael Pollan's.)

In return for food, board and protection, the animals furnished humans with milk, eggs, hides and – finally – meat. It was a mutually beneficial arrangement that, in time, left its genetic mark: the animals became tame, losing the traits that once allowed them to fend for themselves; humans traded the peripatetic life of hunting and gathering for agricultural settlements, and gained, for most of us, the ability to digest lactose.

For the cow as a species if not each individual cow, the alliance with humans has proved remarkably successful: there are now one *billion* domesticated cattle in the world. By contrast, the last auroch, their wild ancestor, became extinct in 1627.

And if, as the animal rightists I lived with that summer assert, it is wrong to exploit animals for human gain, what about dogs? Just about all of the activists owned pets back on land, some of which – coincidentally – were vegans like their owners. Was that another kind of exploitation?

Gollion's cattle weren't pets. My father wouldn't bother to call the vet if a cow was bitten by a snake, but when in 1993 his favourite dog, Marty, *was*, he tried to resuscitate her with an air compressor, shoving the hose down her throat and flicking on the power to try to jump-start her heart. I couldn't tell the cows apart: I knew there were 'leaders' and 'followers' in every herd, but to my untrained eye they all had the same semi-wild aloofness: tame enough to stand their ground when I slowly approached, but not enough to be patted. Cut the fences, and they likely wouldn't have lasted long before they were hit by a car, succumbed to disease or, in a drought year, starved.

That's not to say threats to life ended at the farm gate. During my first calving season I saw cows abandon their calves because of some deficiency imperceptible to the human eye; I watched crows hopping around a lamb that wasn't yet dead but could no longer walk. My parents always did what they could – as children, my sisters and I would awake to find poddy lambs in the tea chest beside the hearth-warming wood stove.

Sometimes the familiarity of this process would transform an anonymous farm animal into a beloved pet. Black Billy (my nephew named it after himself) was an orphaned calf Dad had reared by hand a few springs earlier. When he found it dead from a suspected cardiac arrest brought on by snakebite (BB's frothing mouth was still full of the grass it had been eating), he cried.

But my father never framed the death of a farm animal as defeat or failure on his part; it was just something that happened. What to do when no treatment worked? He always had the same answer: 'Best let nature take its course.'

I came to regard the cattle yards as a kind of agricultural tollbooth. Whenever I found myself in them with an animal, whatever the occasion, I understood a tax would be levied on both parties. A physical toll was extracted from the cattle and a psychological one from us. It was the cost of the financial gain we generated from them. Smacking the cow, I now suspect, was an expression of frustrated misapprehension of this arrangement.

Farmers who wish to keep animals, I was beginning to understand, have no choice but to inflict pain; the only choice they have is what level of pain they are prepared to inflict. *So then don't farm animals*, my anti-whaling acquaintances would say. But I have never been able to reconcile what I consider the fundamental paradox of animal rights: in extending the humanity we enjoy to animals, we are asked to deny our own animality.

*You've been desensitised!* that ship-full of anti-whaling activists would've shouted in unison. Kill enough, the theory goes, and you grow numb to the act. I don't think that's true. Papa won the Military Cross serving as a commando in World War II. He shot humans and was shot by a human; according to family legend, when he was struck in the arm by a Japanese sniper's bullet, he turned to his men and quipped, 'Who did that?'

But after the war, my Aunt Amanda told me, he found it harder to kill *anything* than he once had. When butchering sheep for his family, he would first hit the 'killer' on the head with a hammer to knock it out before slitting its throat, lest the animal catch wind of its fate.

The closest my farmer father got to combat was failing the eyesight test when he was drafted for the Vietnam War. But when I'd asked him, when we slaughtered the sheep, if killing became easier each time, he shook his head. 'I used to slit sheep's throats when I'd slaughter them, but now I shoot them in the head first.' Shooting the wether meant it felt no pain, but there was still the cruelty of taking away its life.

I didn't enjoy hurting animals either, and I'm sure the same is true for most farmers. The frustration and shame of these interactions seemed a more plausible explanation than sheer impatience for the swearing, the punches thrown and the poly-pipes swung in the yards. The animal behaviourist Temple Grandin has written on the importance of rotating abattoir roles regularly to lessen the moral burden on employees of regular killing. 'Nobody should kill animals all the time,' she has said.

Would I ever be comfortable inflicting pain? And if I did, what would that make me? If I was thinking about this already, was I entering the right line of work? All these thoughts stewed around in my head, but I preferred not to think about them. It's a common refrain that consumers should know where their food comes from, but what about knowing the effect the production has on the producer?

It took me months to admit to having assaulted one of Dad's cows. We were walking up the gentle slope from the cattle yards – now dry and dusty, the dappled shade cast by a lone yellow box tree barely lowering the sweltering January heat – after loading a 'broken down' bull and some steers onto a cattle truck for market. Harold, the truck driver, had been, in Dad's words, 'a bit trigger-happy with his cattle prod', and I was prompted to confess my own violent outburst.

My father was unmoved. 'Speaking of heartless bastards,' he replied calmly, 'I caught another possum the other day.' Throughout the summer, Dad explained, he had been setting a feral cat trap in the top orchard each night, baiting it with a piece of fruit he'd plucked from a nearby branch ('that trap and my peaches will catch a possum any day of the week').

Killing possums is illegal – they are a native marsupial, protected under federal legislation – but they are also a frugivore, subsisting on fruit, seeds and nectar. For my father, they were a pest. It was news to me, but he had already trapped and shot a possum earlier in the week. When he found the trap occupied again the next morning, he knew he couldn't shoot another one, he told me, so he let it go. Wouldn't it be back in the orchard come nightfall? 'Most definitely.' But he was content, he said, with his decision: 'I couldn't knock off *two* possums in one week.' Three decades of farming had taught him the precise level of pain he was comfortable inflicting. I was coming to see that not everything I had to learn could be taught.

# 5

When we were kids, my sisters and I weren't allowed to watch TV during dinner. The risk of seeing John Howard was too much for my parents to bear. In the months after he became prime minister in 1996, Mum and Dad wore their opposition proudly, chortling of his imminent demise and slapping a 'Don't blame me, I voted Labor' sticker on our dusty family van.

But as the Howard months became the Howard years, their mood turned first to frustration – Dad no longer referred to him as 'the miserable little man' but as 'the little shit' or 'the little dickhead' – and eventually to censorship. Should the PM slip through their low-fi parental block, unexpectedly popping up on *The 7.30 Report* beside Kerry O'Brien, he could expect an incoming missile.

But I for one relished what televised glimpses I could manage: those bushy eyebrows, suddenly plucked to make him electorally palatable; the chunky bulletproof vest under his shirt after the Port Arthur massacre; the loud shirts at APEC summits; the louder Wallabies tracksuit on his morning powerwalk ... He was a strange sort of fashion icon.

My favourite item in Howard's wardrobe was his Akubra hat. Reserved for visits to marginal, rural electorates, it was always accompanied by a Driza-Bone coat – irrespective of the forecast – and a pair of R.M. Williams boots.

What I found most intriguing about the hat was its pristine condition. Flat-brimmed, symmetrical and impermeable, it was so different from the hats my father wore – tatty, smelly rags of things, stained lurid pink with herbicide dye and full of holes to facilitate melanoma growth.

I turned thirty in the spring of 2014, my first as a farmhand. In honour of this new direction, and largely in jest, three friends gave me an Akubra for my birthday. It was, its label informed me, the 'Riverina' model, one of the largest, deepest and widest styles in the Akubra catalogue, named for the agricultural region in south-western New South Wales where my mother was born, and where Papa once farmed.

It was only when I took it off and studied it – flat-brimmed, symmetrical and impermeable – that I realised what I had been gifted: Howard's hat. In becoming Dad's deputy sheriff, I resembled the Deputy Sheriff. And wearing it made me feel as much of a phoney.

The hat came between father and son, interrupting what I'd naïvely construed as a budding bromance. The thing was hard to ignore: I had to doff it as I entered the ute to avoid hitting my head on the roof, but putting it on the dashboard made visibility difficult.

Dad thought I looked ridiculous, and he said so. 'The bigger the hat,' he quipped on a near-weekly basis, 'the smaller the brain beneath it.'

Because as well as prime ministers courting the pastoral vote, this suspiciously clean, oversized mélange of rabbit pelts was the headgear of hobbyists, of weekend warriors. In New South Wales they are derided as 'Pitt Street farmers' and in Victoria as 'Collins Street farmers'. The difference between them and me was, Pitt Street farmers have high-powered jobs on Pitt Street. They can be hobby farmers because they have *jobs* to support their hobbies.

What was my job? I wasn't a farmhand, because Dad wasn't paying me to help him at Gollion, and by his own definition, Gollion wasn't a real farm. I wasn't an apprentice, because that implied the acquisition of a trade – and the eventual promotion to master. There was already an incumbent farmer; we had not discussed succession. I was more like an unpaid intern: a work experience kid with no guarantee of future employment.

Dad knew I was a writer, but he considered *that* my hobby (I'd overheard him tell a family friend as much), a source of pocket money, maybe a form of intellectual self-improvement. (As well as turning thirty, in September 2014 I published my first book. My father's feedback was that I 'should submit it as a PhD').

That left the two days a week I 'worked' as a university research assistant. Dad's guess at what I did there was as good as mine. I had been employed by an affable professor of politics soon after I graduated, and was promptly given a large, sun-filled office with my name on the door should anyone need to find me. The only person who ever did was the professor, who would apologise for interrupting my frantic minimising of YouTube browsers to tell me what a fine job I was doing. I realise this is more work than many people with office jobs do, but at least they make enough money slacking off to put some aside after paying the rent.

The state of my car was an indication to my father that I wasn't, in his words, 'getting ahead'.

The Akubra kept the sun off my head as spring turned to summer. But what it deflected in ultraviolet rays the hat seemed to magnify in paternal scorn. On our farming days Dad grew snarky, dismissive. He critiqued the way I held a shovel and dug holes, making me feel as horticulturally competent as a visiting head of state awkwardly posing with a tree they'd 'planted' on the south lawn of the White House.

Then there were the pointed comments about my future, grenades he lobbed into moments where silence would've suited me just fine:

'If *I* were you I'd be focusing on your social life rather than hanging around with your old man.'

'Why don't you knock off early today, go and chase some girls …'

'Many of your friends interested in agriculture, are they?'

Worst was when he suggested I should 'chat up' one of the 'lovely Asians' at the university campus where I worked. A Chinese nurse had recently complimented the volume of his chest hair as she shaved it off during a minor medical procedure, propelling all women of Asian heritage to the top of his future-daughter-in-law wish-list. The creepiness of the construction – *lovely Asians* – illustrated why discussing potential mates with my father was to be avoided at all costs. One evening in Canberra a few weeks earlier, a non-farming day, I was eating a kebab alone on a bench when I spied him walking to the community centre where he took German conversational classes. I saw him before he noticed me and shamefully snuck off around the corner lest he raise my sad, dateless dining habits the next time I went to Gollion.

One afternoon we were lying in the dirt, fixing a pump, when he came out with this gem:

'Your friends would be all buying houses by now, I suppose?'

*My* friends? There was Frank, whose LinkedIn account featured a one-word résumé: 'Drifter'. Frank once worked for Greenpeace, chaining himself to bulldozers, but he had been out of work for so long his mother thought him unemployable. He and his partner subsisted off bin-dived bread, home-grown vegetables and roadkill kangaroos. An industrial designer by training, he had recently designed and built a house for his parents using gleaned materials. It cost forty thousand dollars.

Nick? He aspired to Frank's lifestyle but hadn't yet acquired the requisite skills, necessitating employment cleaning toilets for social housing tenants. He bin-dived his groceries, though. (I sometimes joined him after late-night drinking sessions; we'd pre-emptively ward-off hangovers by feasting on low-quality barbecue chickens. Raised on battery-farms, they retailed for just $8.99, yet still ended up in the dumpster. This disgusting expression of industrial agriculture was largely lost on us as we drunkenly tore off drumsticks and wiped our hands on whatever we could reach – trees, grass, the sides of the dumpster itself.) Nick was perhaps the *furthest* from home ownership of my friends: one summer he contemplated living in a tent on Canberra's Mount Ainslie, but when he asked if he could shower in my share-house bathroom, my housemates made me talk him out of it.

Benji? He mainly lived out of his car, a 1984 LandCruiser Troop Carrier with a mattress in the back. He parked it alternately on the nature strip outside my share house and at the university (for the free showers), where he was onto his fourth or fifth degree in an attempt to avoid paying off his student loan.

That left Harry, my oldest friend, and the only one to own a house – and that's because he built it himself. Harry was the scion of agrarian socialists who verged on being 'real' farmers, according to Dad's metric (they produced fine-wool merino on a 3000-acre escarpment). They grew much of their own food and built their own houses. 'It's a bit *peasanty*', Dad said of their lifestyle.

Being a 'peasant' was just about the worst slur in my father's inventory of insults, connoting medieval amounts of mud and deprivation. The term is used ironically by double-income tree-changers these days as a humble-brag ('*We* like to use every part of the pig, from snout to tail'), but my father was not in possession of an Instagram account. Necessity vs lifestyle: in my father's eyes, that's what separated the hobbyists from the hicks.

Things came to a head at the end of 2014, a few months since I'd begun helping on the farm. My mother had read that an upcoming bull auction was being held a few gates down from the farm she grew up on in western Victoria; I met her and Dad in Melbourne the morning of the sale, and we drove west across the flat volcanic basalt plains.

At the farm where my mother grew up, the current owner didn't mind us dropping in unannounced, and was happy to down his welding tools to give us a tour. But it was Mum's tour, not his: she wanted to show me Koonje, the farm as it had been when she knew it (it was now called 'Sunny Ride' or 'Sunnyside' or some such): the rows of native seedlings Papa planted as wildlife corridors that were now fully grown ('he was a very forward-thinking farmer'); the windmills he had set up (he had been chastised by Ma for using his handkerchief to check their oil); the old hay paddock where Mum and her siblings had 'tanned' themselves, with predictable results for the pasty children of Anglo-Saxon stock. She was overcome by it all, and we posed for photos to show her siblings.

It was late spring when we visited. Spring was a noisy time at Gollion: the carols and shrieks of willy wagtails and noisy friar birds; the rat-a-tat-tat of restless flycatchers and baleful cries of baby magpies. But here all I could hear was the distant rumble of a tractor ploughing dark, moist soil. It was all so flat, featureless and productive; so *foreign* to me, having grown up in hilly, marginal, drought-prone Gollion. In so much of Australia, the early settlers had tried to recreate England with predictable failure; here they seemed to have succeeded.

To Dad's annoyance, by the time we left Koonje we had missed much of the bull sale; his mood deteriorated further when, on arriving at the sale, I produced my Akubra from my suitcase and put it on my head.

On the drive back to Melbourne, I remembered I hadn't paid my rent. I transferred the money on my phone, then realised I had nothing left over until next pay day. I asked my mother for a loan of petty cash; Dad overheard. 'Why don't you get a *real* job so you don't have to leech off your mother anymore?'

I asked to be let out and slammed the car door, my Akubra casting long shadows across the streetscape as I stormed off to find the nearest train station.

Then, one morning a few weeks later, an olive branch. At the very kitchen table where I had hatched my plan to become a farmhand four months earlier, my mother suggested something that would change the course of my life. I'd been working so hard helping my father, she said; it might be time to start a farming project of my own.

Dad had already attached the ripper to his tractor.

PART II

# Orchardist

Every plant, every crop that a person tends, bestows its qualities. For example, it's good to cultivate ficus. True, ficus is a fruitless plant, but it compensates by giving something of itself. Whoever wants to get rich, let them cultivate ficus. It brings peace and well-being. Ficus is for the learned.

**Beinsa Douno**

# 6

I needed a crop. But what to plant? If this half-acre hillside I had been granted for my project was simply a way for me to increase my sense of proprietorship – a patch of my own to tend on my farmhand days – then what I grew was immaterial. But there were other considerations. The Australian summer of 2014/15, when I got around to fencing off the ripper lines, was, like the summer before it, and the summer before that, among the hottest on record.

Water wasn't a problem. Gollion's gardens and existing orchards were irrigated from a series of dams that flowed downhill into each other. The water – brown and cold, sucked from the depths – was pumped, with the aid of gravity and a high-pressure siphon, 500 metres from a dam that sat at a higher altitude than the orchards and garden it would irrigate.

Heat was a problem. No amount of water can alleviate the stress on plants of 40-degree weather, once a rare occurrence but now an inevitable part of every summer. (I remember the excitement in primary school of being sent home early because the mercury had hit the magic number.) That ruled out several species of fruit trees once suited to south-eastern Australia but increasingly reduced to ornamental status as harvests withered and dropped during heatwaves: cherries, apples, pears, hazelnuts (all of which my parents had planted).

I wanted whatever I grew to be unique, at least locally. Olives blanketed several hillsides on neighbouring farms, so were out. Gollion fell within the Canberra cool climate wine district, but there were already over one hundred local vineyards, some of which were already removing Pinot Noir and Riesling vines as that 'cool' climate heated up. Also: I knew nothing of winemaking.

And, I wanted to make some money. This ruled out most temperate fruit, which supermarkets have conditioned customers and chefs into believing aren't worth more than a few dollars a kilo, including those that thrive in the heat (apricots, nectarines). Some fruits were almost anti-capitalist in their unsuitability for retail (white peaches are my favourite-tasting of Gollion's fruit, but are so delicate they bruise by the time they make it back to the house; Cox's orange pippin apples are delicious and rare but ugly little things, covered in what looks like a patina of rust).

Truffles were suggested by friends and family. This part of Australia, with its cold winters and dry summers, is, along with parts of south-west Western Australia and northern Tasmania, one of the best places to grow them in the country, with growers commanding high prices as off-season suppliers to the European market.

Harvests can come in as little as four years, but first you have to inoculate oak or hazelnut trees with the subterranean fungus that produces truffles and, in the acidic soils of the Yass Valley, spend time and money trucking in tonnes of lime to alter the pH (50 tonnes for each hectare is the recommended ratio).

That was another constraint: time. I didn't want to wait decades for a saleable product. Pomegranates ticked the boxes for climate, price and gap-in-the-market, but although they bear fruit after a few years, each crop needs up to seven months on the branch to mature – seven months when they would be exposed to heatwaves, storms, hail. I considered persimmons, but then I read that they are sensitive to wind. The site of my new orchard, below the existing lower orchard, was a very windy place. I even considered using the space to produce something other than fruit. Hemp? Meat chickens?

In the end, I decided to grow something with a proven record. A crop I knew thrived at Gollion, was not grown locally at a commercial scale, and loved hot, dry summers. One that produced saleable fruit within a few years, and which people associated with luxury. I only needed to walk through the top orchard for my answer.

By the end of 2014, the original orchard was thirty-one years old, an arboretum of most of the temperate fruit trees that grow in Australia. Creaking open the gate and ducking under the first of several neck-high irrigation pipes, you immediately noticed the sheer abundance of *life*. Not just the trees themselves – glossy, canoe-shaped leaves on the peach trees nearest the gate; red-green fuzzy teardrops on the adjoining pears – but *in, between* and *below* them. There were bugs on the branches and spiderwebs across the rows (always, I noted, at face-height). Butterflies fluttered. Crickets trilled. The sward of grass contained species not common elsewhere on the farm and was largely left unmown in summer to provide a refuge for pollinators and other 'good' insects that, my father explained, preyed on mites and parasites afflicting the trees.

When, Dad told me, this orchard first produced fruit ('a few little peaches'), he took a handful to Jack Whyte his neighbour at Westmead Park. The old cocky's response: 'You didn't grow them on that *rocky hill*, did you?' Now that rocky hill was a lush microcosm of what regeneration looked like.

At times there was too much life, which was why a kennel sat in the dense shade of two cherry trees, from where Suey Dog (in theory at least) would scare off brushtail possums on summer nights. There were holes in the roof of the net made by trees trying to get out and birds trying to get in, as the regular blue-and-red flash of crimson rosellas attested. Suey's predecessor, Harry Dog, was buried under a young quince tree where he once liked to feast on the birds, which he caught with athletic aerial pirouettes.

The orchard, like the rest of the farm, was organic but not certified so, a process that costs money to the producer, commands a premium from the consumer, and is regulated by annual audit. (When you see a stall at a farmers' market describing themselves as 'in transition' to organic, it means they have stopped using chemicals, but have not yet reached the three years of chemical-free farming that qualifies a grower for certification; I have since come to tell chefs that our fruit is *de facto* organic, not *de jure*).

My father hadn't bothered with organic certification. This wasn't a *real* farm, after all, let alone a commercial orchard, so we didn't need to convince anyone of its worth.

Not that bureaucratic oversight has ever been his style ('you don't want blokes with clipboards nosing around your place'). And anyway, the occasional use of herbicides to control serrated tussock on Gollion's southern boundary disqualified us. 'Biological' pest control came instead in the form of a tunnel connecting the chook shed to the orchard (the hens ate caterpillars and turned them into fertiliser before they could turn into moths), and in the stands of grasses and their bugs, enhanced by gone-to-seed parsley, carrots, celery and fennel in Mum's adjacent vegetable garden. These plants, she told me, when left unharvested produce 'umbels', umbrella-like flowers that radiate from a single point. They are helipads for aphid-eaters like lacewings and ladybirds.

To attract the other beneficial insects, the pollinators, the gardens around the orchard contained nectar-rich daisies that flowered at different times of the year: marigolds, chamomile, marguerites, sunflowers. And, as well as being the principal ingredient of Dad's compost and fodder for cattle, fast-growing tagasaste or tree lucerne produced masses of drooping white flowers that helped attract bees each spring. A plantation of them spanned the top orchard on three sides, in spring creating the low-frequency hum of a distant sporting crowd. Once again, what I'd assumed was farmyard messiness was actually environmental order, a considered plan of action; a way to commandeer nature into doing some farm work.

For ailments that couldn't be eaten by others – fungus and blight, virus and disease – Dad hauled a backpack spray unit across his shoulders and blasted the affected leaves in a mist of whatever remedies 'the greenies let you get away with' (copper oxide, lime sulphate, Neem oil) with an enthusiasm that evoked, when combined with his overalls and Bill Murray hairline, a retired Ghostbuster in search of a ghost.

Zigzagging across the orchard's rows, from the bottom to the top, you saw other, less successful attempts at pest control: a fake metal crow dangling from a wire; glinting CDs, twine looped through their holes and hung from branches, a novel if ineffective way to recycle yesterday's technology. In an organic orchard there is always something under attack, and over the decades trees had died, been removed and new ones planted in their place; I don't think anyone knew anymore how many cultivars grew here, or even what they all were.

Continuing across and up the hill, ducking under the irrigation lines and pulling spiderwebs off your face, you saw hedges of hazelnuts and bushes of pomegranates, rough bark on pear trees and smooth bark on cherry trees. And then, towards the south-western corner, in the third row from the top, stood the widest, healthiest and brightest of all the orchard's trees. Barely pruned in three decades, the shape and colour of a giant mould of lime-flavoured Aeroplane jelly: *Ficus carica*, the fig tree.

The fig is not a fruit but a flower, the hidden pearl of a closed oyster. Its jammy core comprises a series of tiny blossoms; the crunch they produce, seeds. There are more than seven hundred and fifty species in the *ficus* genus, and they come as shrubs, vines or trees. They grow in deserts and in the tropics and are considered a keystone species by scientists, who plant them first when regenerating rainforests because they attract so many different feeders. Fig trees grow in disused railway yards and on abandoned rooftops;

on cooled lava fields and sheer rockfaces. In Cyprus, in 2018, a fig tree was found growing in a cave, having sprouted from a man's stomach when he died there in 1974.

Romulus and Remus were found being suckled by a she-wolf in the shade of a Roman fig tree. Buddha attained enlightenment under one and Jesus threw a tantrum at another when it bore him no fruit. Figs are the most mentioned fruit in the bible, and surah 95 of the Koran opens with an oath to them. Vulva-shaped figs – the Greek word *sykon* means both – convey promiscuity, but their loincloth-leaves protect the modesty of parkland statues. Figs can symbolise arrogance (the acronym FIGJAM stands for 'Fuck I'm Good Just Ask Me'), nonchalance ('give a [flying] fig') and hopelessness (ancient Hindu scripts wrote not of seeking needles in haystacks, but flowers in fig trees). Figs may have been the forbidden fruit eaten by Adam and Eve (the Holy Land is too hot for apples), and Fig Newtons are surely the last things eaten by many residents of nursing homes. They convey luxury (consider their early-season price) and vulgarity (to 'give the fig' by wedging your thumb between the next two fingers is in many countries the equivalent of flipping the bird; Vanni Fucci gave God the fig in Dante's *Inferno*).

And, whether due to the soil or the climate or the aspect or its genetics, Gollion's black Genoa fig tree was special. From late February until early May it produced a steady crop of purple-skinned, red-fleshed figs, shaped like Chinese steamed buns and tasting of the sun. Figs don't ripen off the tree, so my parents waited as long as possible before harvesting them; the fruits' skin cracked and an ambrosial teardrop of syrup formed at the end of each one by the time they were cut free with a knife. ('What fruit has the eye of a widow and the cloak of a beggar?' asks an old Spanish riddle.)

I knew they had commercial potential from the reaction of those who tasted them. My parents had never sold their fruit; they gave the masses of surplus harvest away to friends and family. But they did barter some of their figs for bread with Canberra's best bakery. (A good deal for the bakery: boxes of figs that could be thinly sliced for pastry toppings, all for the price of a few loaves of sourdough).

When I started working at Gollion I started charging the bakery (I based my price on the retail markets in Sydney) and took samples of the figs to friends who worked in kitchens to see if they'd like to buy them too. Without exception, they did.

Louis Couttoupes, a giant man running a tiny kitchen, banged his head on the exhaust canopy the first time I plonked a full box before him, a scene that likely repeated itself often, but one I took as a sign of his excited approval. Sean McConnell, head chef of Monster Kitchen and Bar, declared them the best figs he'd tasted and was soon buying as many as I could supply. His pastry chef, a normally friendly Savoyard named Angélique Peretto, would become possessive when I walked in, clutching each delivery of figs to her chest and warding off colleagues with a rolling pin if they tried to appropriate some for other stations of the kitchen.

These were gifted chefs, discerning about the produce they sourced. In 2017, Louis's employer, Bar Rochford, would be named Australia's Best Bar by *Gourmet Traveller* magazine. Sean is the brother of Andrew and Matt McConnell, patricians of the Melbourne restaurant world; between them the three siblings have accrued seven hats from Australia's *Good Food Guide*. A couple of years after arriving in Australia, Angélique was offered her own cooking show after she scored a perfect thirty points for her green tea and coconut mousse, topped with raspberries and wrapped in a vanilla tuile, as a guest contestant on *MasterChef Australia*. Angélique, who finds much of Australia's dining scene vulgar and superficial, declined the offer, explaining to a surprised producer that she wanted to maintain her *intégrité*.

All of them were excited when I told them I was planting an entire orchard of black Genoa figs. They knew what was good. Didn't they?

It was at the Hobart Farm Gate Market, among the activewear and the granny trolleys, that I got talking to an American baker. I was visiting a friend, himself a baker, an amateur but very good one, who had learned some things from this professional, a master of natural-leavened breads and Bay Area sourdough starters.

I bought a croissant, and as the professional baker – bags under his eyes, all those early mornings – pulled it out of a basket and popped it in a paper bag, I studied the fruit pastries: apple, pear, plum. Perhaps I studied them a little too intently. The amateur baker introduced me as a visiting friend, one interested in growing fruit; he said I was establishing a fig orchard.

The professional baker handed me my change. 'You can't grow good figs here.'

Oh no, I explained, the silly mainlander not making

himself clear, I don't mean Tasmania; where I live, near Canberra, it's hot and dry, just the right conditions for figs.

The professional baker guffawed. That's what he did: guffaw. I don't think I'd been the subject of a guffaw before. I didn't like it. 'Australians don't know what good figs taste like,' he said curtly, 'because you can't grow good figs in this country.'

I protested that, for thirty years, Gollion's tree had produced thousands – perhaps tens of thousands – of *very* good figs. I had in fact eaten several hundred myself, so could attest to both their existence and excellence.

'Ours have a lot of fans,' I said, 'including some really great chef—'

'That's because they haven't tasted a *real* fig. You need fig wasps to produce real figs. The wasp enters the fig and pollinates it. Without the wasp, your trees won't produce pollinated figs. And fig wasps don't live in Australia. It's common knowledge. Aristotle even wrote about it.'

Fig *wasps*? I was defensive, embarrassed, confused and beginning to have the awful feeling that perhaps this professional baker (the croissant was indeed very good) was right.

'Whatever,' I said.

'Yeah, *whatever*,' he said, and turned to serve his next customer.

I consulted Aristotle: the dickhead baker was right. In his *History of Animals* (c. 350 BC), the Greek ur-zoologist wrote:

> The fruit of the wild fig contains the psene, or fig wasp. This creature is a grub at first, but in due time the husk peels off and the psene leaves the husk behind it and flies

> away, and enters into the fruit of the fig tree through its orifice, and causes the fruit not to drop off; and with a view to this phenomenon, country folk are in the habit of tying wild figs on to fig trees, and of planting wild fig trees near domesticated ones.

Wild figs? Domesticated ones? Country folk?

I read some more. A few decades after Aristotle's *History of Animals*, his one-time colleague and philosophical successor Theophrastus identified the male wasp, which never exits the inside of the fig, and is 'sluggish like a drone'. Pliny wrote of the wasps, as did Herodotus (he called them 'gadflies'). The figs for sale in Australia are native to the eastern Mediterranean; they were an important food of antiquity and clearly worthy of scientific inquiry.

The process, I now know, is called caprification. Because fig flowers bloom internally, they can't rely on wind, bees or birds for pollination. The figs eaten by Aristotle were the progeny of two trees, the 'wild' (male) fig, and the 'domesticated' female. Fig wasps need figs to hatch their larvae, and fig trees need the fig wasp (*Blastophaga psenes*) to spread their genetic material. This symbiotic relationship starts when the female wasp, a tiny creature with wings as wide as its body is long, crawls into the ostiole (the small opening at the bottom of every fig) on the male tree and lays her eggs. The figs of the male tree are hard, small things (they are called caprifigs, from the Italian word for goat, the only creature they were thought suitable for), with hundreds of internal hollows in which the larvae can pupate. The female wasp's wide wings snap off as she crawls into the caprifig, where she is now trapped and dies. The eggs hatch. The male wasp, born without wings, will die inside the fig as well – but not before he has mated with his newly hatched sisters and tunnelled a way out of the fig for them.

Freed from the fig, the female wasps fly off to lay eggs of their own, carrying with them pollen from the male fig tree on their backs. If they accidentally enter the ostiole of a fig on a female tree rather than a male one, their wings will break off and they will die without finding enough space to lay their eggs inside. But the pollen they have brought with them will fertilise the blossom of the female fig, allowing the 'fruit' to form. One blog I read claimed, 'According to some fig connoisseurs, pollination produces a more delicious fig with a superior nutty flavour due to the seeds.'

Our figs didn't have a nutty flavour. Because they produced no seeds? They looked like seeds. Were they unpollinated? Where were the wasps?

The internet is full of the wonders of caprification, mostly written by people squeamish at the prospect of eating a dead wasp in their supermarket fig. I seemed to be the last person – certainly the last prospective fig farmer – to know. 'It's common knowledge,' the professional baker had assured me. 'When you eat a dried fig,' a *New Yorker* article from 2016 explained, 'you're probably chewing fig-wasp mummies, too.'

But there was a correction at the bottom of that article: *This article has been revised to clarify the fact that not all fig plants require pollination to produce edible fruit.* (Another correction, never made, is that any 'wasp mummies' would be broken down by the fig into proteins by an enzyme called ficin – it's not like you'll find a perfectly preserved wasp if you cut a fig in two.)

So, there were figs that required wasps to fruit, and figs that didn't. I kept reading. The former camp is called caducous or Smyrna figs, named for the region in western Anatolia where they are from. As well as giving them a supposedly 'superior nutty flavor', pollination increases the sugar content of these figs, which makes them suitable for drying. (I have since noted the provenance of every packet of dried figs I come across. With some Iranian exceptions, all have been Turkish.)

The second type of figs are called persistent or common figs. They are self-pollinating, contain sterile seeds and come from trees that are only female. Likely derived from a mutation that allowed the growth of fruit without wasps, they must be propagated deliberately by cuttings.

Persistent fig cultivars, grown to be consumed fresh, include black Genoas, brown Turkeys, and the white (actually green) Adriatic – by far the most common trio available in Australia. The American baker's insinuation: that in thinking these varieties were good, Australians didn't know what we were missing.

But we once did. A 1911 newspaper article from Perth's *Western Mail* confirmed that Smyrna figs had grown in Western Australia since 'the early days of the State' (probably from the seeds of dried figs). Without anyone knowing why, the trees never produced figs.

In South Australia and Victoria, the article continued, 'certain enterprising citizens' deliberately planted Smyrnas at the turn of the twentieth century. There, too, they had refused to bear.

It took years for anyone in Australia to realise wasps were needed to pollinate the Smyrnas, and in 1908 the country's first crop was produced by one T.B. Robson in his South Australian orchard, with wasps introduced from Cape Town.

Over the next few years, more farmers experimented, and in 1911 Mr J. Hawter exhibited two dishes of figs at the Royal Perth Show. For fig farmers, it was the equivalent of the atom being split (SMYRNA FIG IN WESTERN AUSTRALIA. A FUTURE INDUSTRY. MR. J. HAWTER'S SUCCESS, read the headline). Fertilised figs, the author wrote, 'develop into large beautiful luscious greenish-yellow fruit, very rich in sugar, which dries readily and produces the celebrated figs of commerce ... its delicate colour. Its thin skin and its exceedingly fine flavour. No fig I have tasted in Western Australia can compare with it in this respect.'

The article ended with advice for growers wanting to get into the commercial fig game:

> I would give one caution very emphatically, and that is this ... fig business is not one for the man who is struggling with limited capital. In such cases one cannot too strongly recommend the fruit grower to confine his attention to those things which are proved, and which are most certain to give profit. While I strongly believe in the future of the fig industry, it is equally necessary to acknowledge the fact, that my opinions have yet to be proved.

Sometime during the following century, the practice of caprification vanished from Australian agriculture. In a fact sheet from the New South Wales Department of Agriculture, I read that Smyrna figs had once been grown for dried fruit production but 'declined in the 1950s, when imported figs began to dominate the market.' What became of Western Australia's optimism, or South Australia's earlier success? I found nothing.

I became preoccupied with what I have come to consider The Wasp Question. If wasp-pollinated figs really were superior, why were they no longer grown in Australia? Or were they? What if there were still Smyrna and caprifig trees growing – forgotten, neglected, *feral* – in Australia? Theoretically, if a grove of caprifigs existed and was supporting a population of wasps, could I plant some at Gollion and somehow tempt the wasps to move in?

I emailed Steve Compton, an entomologist and expert in fig wasps at the University of Leeds in England. I told him about my exchange with the American dickhead in Hobart but replaced the word 'dickhead' with 'friend', just in case.

Compton didn't think I or my 'American friend' need worry: taste differences are 'just varietal variation, independent of whether the variety needs wasps or not.'

*Just* varietal variation? I still wasn't sure if I was choosing the best-tasting fig.

Compton was unsure if there were still Smyrna figs in Australia but explained that, hypothetically, the wasps can travel long distances '(>100 km)'. If I planted a small number of male trees at Gollion, the wasps should be able to find any female trees still present in the region. 'It will be a major project to get them and the wasps established though.'

Next I emailed Finn Kjellberg of the French National Centre for Scientific Research, according to Steve Compton 'the world authority on *Ficus carica*'. Kjellberg agreed: 'If you really want to have pollinated figs (and I think you don't) then you need to be in a sufficiently seasonal climate for the yearly life cycle of the fig wasp and the yearly fruiting of the figs to work out right.' Kjellberg didn't know if there were still Smyrna trees in Australia either. As for whether or not they were superior, he said it depended whether you wanted dried figs, which need a higher sugar content, or fresh ones ('the advantage of pollination is that figs get sweeter. The disadvantage is that the figs get sweeter').

I kept looking, but I couldn't find any evidence of introduced fig wasps still living in Australia. It seems that with improvements in freight, storage and packing, Australian farmers couldn't compete with cheaper and bigger growers overseas, especially in Turkey (which still produces the majority of dried and fresh figs). The fate of Smyrna figs and their wasps now strikes me as symbolic of our wider paucity of food diversity in Australia – a preference for produce that travels easily, keeps well and looks good over weird and ugly heirlooms, even if the heirlooms often offer better nutrition and flavour. There are around 250,000 flowering plants on Earth, many of which we can eat. Yet half of the calories now eaten by humans come from three plants: wheat, rice and corn.

Variety is the spice of life, and there is strength in diversity (strength against disease, malnutrition, crop failure). But convenience and efficiency are stronger drivers of the modern food system, as witnessed in pre-sliced apples and small bananas being marketed at my local supermarket as 'kid-sized'. English and American orchardists used to value late-season heirloom apples that wouldn't have to remain stored in a cellar during winter for as long as early-season apples. Now shoppers in those countries can buy the same apples all year round. The off-season ones are imported from Australia.

My attempts to grow waspless figs were proving hard enough. In the spring of 2014 I took thirty twenty-centimetre cuttings from the black Genoa tree in the upper orchard. I dipped the bottoms of each one in honey (to stimulate root growth) and planted them in polystyrene boxes filled with potting mix. I watered them through the summer, and although the tips and the wood stayed green (I checked the wood by scratching it with my fingernail, on Dad's advice. Brown scratch = dead, toss the body on the compost heap. Green scratch = don't give up hope just yet), no buds burst.

In the spring of 2015, I tried again. This time I took fifty cuttings and planted them in a wicking bed, a raised garden where moisture is 'wicked' from below, not watered from the top. This time, they took. Then, in the autumn of 2016, when these cuttings had lost their leaves, I carefully dug each of them up and transferred them to the ground. They weren't much to look at, and I was taking a risk: although nothing attacked the top orchard's fig tree (apart from birds and possums after the actual fruit), in replicating its genetics exactly, I was creating a monoculture, liable to pests and disease. If one tree was affected, they all would be. But they were mine, and they were alive. The next time I did the scratch test, in the middle of what turned out to be a very wet year, each cutting was still green. By November, they all had leaves.

# 7

My father, too, was excited by the fig orchard. A new orchard meant a new project: even if it was *my* project, it needed skills that he possessed and I didn't, which meant new opportunities to impart knowledge and perform tutorials, and new reasons to Get Stuff Done. He assumed the role of a building site's foreman, encouraging me to do as much of the work as I could, but prepared to step in and lend a hand where needed.

I sought his counsel from the start. When I'd asked him how deep to dig my holes for the transplanted figs, he smiled his mirthful smile, revealing the snaggle tooth at the corner of his mouth, and asked if I remembered a colleague of his from his economics days, a man with an Italian surname. I said I did.

'Well, his dad came from the citrus groves of Sicily, which are controlled by the mafia. The old man said that over there, they dig the same hole for fruit trees as they dig for burying bodies – the same people did both jobs!' (In other words: the deeper the better.)

The plan was to extend the netting of the orchard immediately to the west of the new fig orchard, creating one giant structure enclosing what we called 'the teenagers' (fifty trees of stone fruit, apples, quinces and black Genoas planted around fifteen years before) and the entirety of my new orchard. Unlike the top orchard on the other side of the house, which was protected from gales by the shape of the hill and by mature stands of trees, this one was exposed to the winter southerlies and summer easterlies. Essentially, we were building a giant windsock.

The skeleton of poles and wires would need to be carefully engineered to avoid (at best) creating weak spots where the net would tear and (at worst) blowing away. While I unspooled plastic irrigation pipes, Dad drove to and from hardware stores and timber yards, returning with wooden posts (for the outside), smaller steel posts (for the inside), wire, screws and bags of cement.

We resumed the easy male companionship that comes with shared noncommunicative activity. Dad showed me the 'right' way to dig a post hole, by making ever-expanding concentric circles with the blade of a crowbar, before 'having a spell' while I shovelled out the loose earth. (The 'wrong' way, which he had caught me doing, is to start the crowbar circles on the outline of where you want the hole, rather than at its centre. The wrong way makes shovelling harder and results in a convex hole.) Spearing the metal crowbar into the earth when not in use: right way. Laying it down in the grass, maximising its exposure to the sun and so ensuring that when I picked it up again it would burn the blisters that had only recently formed on my prissy writer's hands: wrong way. He showed me the best consistency for concrete (it shouldn't splash you in the face as you pour it into the post hole) and how to strengthen the mix by tamping it until no bubbles appeared (he used his fingertips; I used a stick).

Car tyres were collected and placed around each tree, which helped to keep the irrigation lines from writhing like snakes when the water was turned on, but also stopped rabbits from eating the seedlings and created a humid microclimate for each plant. There were hundreds of tyres scattered around the farm, mostly looped like quoits around previously planted trees – some were now impossible to remove, the trees having grown to maturity inside them. When I asked Dad where he'd learned this technique, he said he couldn't remember, only that he'd bought them second-hand from mechanic yards thirty years ago. Back then, he said, 'when I wasn't as ecologically enlightened as I am today', he'd thrown some excess tyres onto a bonfire. 'You should've seen the plume of black smoke: it was like something from the seven o'clock news; a riot in the streets of Beirut or Paris . . .'

He had told me that on a farm, where there was no end to tasks needing to be done, he stopped himself feeling overwhelmed by making sure he achieved at least one job he'd set himself each day. He took stock of whatever it was before 'knocking off' and retiring inside.

Dad derived satisfaction, I saw, from making something from nothing. Building, growing, *creating.* What satisfied him less? Eating those creations.

It wasn't that he avoided the produce he grew. Self-sufficiency was the reason he and Mum gave whenever asked why they'd planted all these fruit trees. But the longer I worked with him, the more I suspected that this back-to-the-land story was an excuse for planting and grafting as many different varieties of tree as he could, many of which, including the figs, he had no intention of eating.

He was a farmer, not a foodie. Growing a healthy crop in a healthy way was what he enjoyed – it was a challenge for himself, for the climate, and for the competition ('ever *bought* peaches as good as mine, Sam?'). What happened to the crop once it was picked was of secondary importance. For my father, eating was about refuelling. Whether he was filling his tank with homegrown cherries or store-bought Cherry Ripes (their empty packets strewn about his ute, where Mum, much more into the homesteading ideal, couldn't find them) barely mattered.

It was the economist in him. Slaughtering and eating his own lambs may have been culturally ingrained from a young age (or perhaps he'd just crunched the numbers and found that it saved money), but when I suggested we slaughter and eat an old hen that had stopped laying he scoffed and asked me why we'd bother. For him – strangely, for someone who took such pride in growing organic, high-quality produce – spending all day plucking a hen and making chicken stock was a waste of time when you could buy the finished product at Gungahlin Aldi. Matters of taste, nutrition and provenance took a back seat to questions of efficiency, time and money.

It was the unreconstructed bloke in him. Planning and cooking a meal were not in his skill set. To Dad, the fridge and pantry were mysteriously self-replenishing, like the cantankerous magic pudding in the story he'd read us at bedtime as kids.

'You going to the biscuit shop today, Sam?'

'We're out of peanut butter if anyone's going to the nut shop ...'

'Can someone go to the yoghurt shop today? We're out.'

The 'yoghurt shop' I especially liked the sound of, with its Vermeer imagery of maids in tri-pointed bonnets carefully ladling cultured milk from a fire-heated cauldron into clay pots turned on a wheel by the ceramicist next door.

One day we were having lunch, and Dad picked up Mum's *Organic Gardener* magazine. It wasn't the cover image of a handful of onions, dirt still attached to their roots, that perked his interest, but the headline 'KEVIN MCCLOUD ON SLOW LIVING'. Dad flipped the pages as he chomped his sandwich, then called out to my mum.

'Listen to this, Jane: Kevin from *Grand Designs* says, "I'm a big advocate of making and doing – whether that's repairing a chair, or cooking, or knitting a cushion cover, anything that doesn't involve looking at screens. These things put us in touch with a quarter of a million years of evolution. Physiologically we are not designed to sit and look at screens. We are designed to amble and walk, to collect berries, to weave clothes and make furniture, to talk and socialise, to be part of a community. When we do these things, they are incredibly satisfying and are great stress relievers. We produce a lot of serotonin when we do these things as opposed to [here Dad paused, unsure of how to pronounce a word he hadn't encountered before] ... *dop-a-MEEN*, which is what you produce when you go shopping."

'Jane, what's a *dop-a-MEEN*?'

His wife's correction came from the next room.

'*Dopamine*. Never heard of it. I guess 'cause I never go shopping!'

But that wasn't entirely true. Like many baby boomers who could afford better, my father was in love with the

German discount supermarket chain Aldi. Why bother picking leafy greens from the garden when you can buy an iceberg lettuce from Gelsenkirchen? Aldi was a source of bargains, surprise (he loved the 'lucky dip' nature of what you'd find on its shelves) and entertainment (the Aldi specials catalogue was his preferred periodical). I asked him once why his love of shopping at Aldi didn't extend to shopping in general, or even just shopping at other discount outlets like Costco.

'Aldi squares with my personality. It's daggy. Costco's too American-y, too consumption-y. Plus at Aldi you form a personal bond; you get to know the staff.' Starved of friends on the farm, Dad had mistaken customer service for friendship.

A few days after he learned the D-word, I heard this exchange between my parents.

Dad: 'You going shopping this morning? Maybe I should come – get some *dop-a-meen*.'

Mum: 'Dopamine.'

Dad: 'Yeah.'

Mum: 'No, David, that wouldn't work for you – you don't get a dopamine hit from buying a coffee, because you're too busy grumbling about the price.'

Dad: 'What can I get dopamine from, then?'

Mum: 'Making compost, probably.'

Modern farmers in rich countries like Australia are no longer all-round growers of food: they are *beef farmers* or *apple farmers* or, increasingly, 'producers': specialists in a specialised economy. Previously, farms grew all they could to sustain their residents, with the surplus sold to market; now the ratio has flipped, with most of what they produce sold to *the* market (a concept rather than a place).

For thirty years my father had been a free-trade economist, and as with farming, he traded in aphorisms from that time. *Trade builds wealth*; *nations, like people, should find their niche*; *subsidies for farmers keep farmers poor*. This last one – uttered in contempt when he saw TV footage of EU smallholders 'with one or two cows' driving brand-new John Deere tractors, or Australian pastoralists asking for handouts to farmland that was in drought more often than it wasn't – translated to a belief that farmers would often be *better off* – richer, happier, healthier – if the invisible hand of the market were allowed to put the unviable ones out of their misery.

In 1901, 14 per cent of Australia's population worked in agriculture. The number would have been higher had Aboriginal and South Sea Islander people, who commonly worked as stockmen and station hands, been counted in the census. According to the 2016 census, which did include Aboriginal people, 2 per cent of Australians worked in Agriculture. There are far fewer farmers in Australia today than there were one hundred years ago, although the overall population has increased by a factor of five. Those farms that remain are bigger, more productive and more specialised and are more likely to be owned by a business consortium than a family. I take Dad's main point: Australia is a much richer nation than it was a century ago, with a service economy and non-agricultural export sector that far out-earns agriculture. But that increased purchasing power has brought deeper inequality and a new kind of poverty.

Compare the relationship my family had to food grown on our farm to that of the Whyte family at nearby Fernleigh as late as the 1990s. We grew our own fruit and some vegetables (but not many, except in summer). We had a few hens, which produced one or two eggs a day. The Whytes – hicks, derided by my father and, unquestioningly, by me – milked a cow each day for fresh dairy products, kept a pig for Christmas, and raised geese, ducks and chickens for eggs and meat. Their diet – complete with a still for moonshine – may not have done much to ward off scurvy, but it is now suddenly very trendy, the kind of menu a hip country restaurant would serve to well-paid urbanites. The Whytes were peasants, but peasants know where their food comes from.

In Australia, where a duopoly of supermarket giants strangles independent retailers, dairy and charcuterie from small, family-run farms are often easier to find in inner-city specialty delis than in the very regions where those foods were produced. The urbanised nature of Australia, where most of the population lives in six capital cities, exacerbates the effect: produce is freighted from farming regions to depots in the cities, where it is collated, with the best quality produce delivered to citywide retailers, before the rest is shipped back to the regions whence they came.

As well as wasting energy through excess transportation, this system forces prices down, with farmers forced to accept whatever price the supermarket giants will pay. Cheap prices encourage farmers to cut corners and take ecological and financial risks for diminishing returns. Even if they have a vegetable garden, they barely have time to tend it.

Over the past two hundred years, agriculture has industrialised in step with the rest of the economy. It has been transformed from a self-contained system, driven by the sun, fertilised by manure, and tilled, ploughed, picked and harvested by many hands, to one that is now dependent on mechanical labour and chemical inputs, producing products determined by the market more than the season.

Throughout the developed world and, since the green revolution, the developing world as well, industrial agriculture has become the dominant model of food production. Decisions are increasingly made not by family businesses but by transnational organisations in the fields of genetic engineering, pharmaceuticals, chemicals, food manufacturing and global commodity trading, enabled by governments.

First in Europe, North America and Australia and latterly Asia, South America and Africa, smallholder traditions of agriculture dependent on traditional knowledge and sustainable land management, on community craftsmanship and labour, and on ecological inputs and the recycling of nutrients have been eroded. Undoubtedly, gains have been made: education and family planning have given options to people who may not want to be farmers where they once had no choice.

But farming for cash crops instead of subsistence has resulted in the perverse outcome that many farmers eat like shit. It all leads to the tragic reality that farmers are among the most alienated people from food. Obesity, heart disease, diabetes – the 'modern' pathologies of our era – are all more prevalent in rural North America, Australia and Europe than in the cities of these continents.

This isn't new. It figures in a 1985 documentary, *God's Country* by Louis Malle, in a scene of a stovetop in rural Minnesota, on a dairy and corn farm run by Jim and Bev Mackenthun. Jim has spent the morning sitting in his tractor, tilling soil the colour of an Oreo cookie, left hand on the wheel, right elbow out the window, 'the easiest job on the farm, I'd say.' He comes into the kitchen, all Formica and linoleum, where his wife is spraying an aerosol can of cooking oil onto a casserole dish and scooping into it the beginnings of a tuna salad. Malle narrates: 'Meanwhile, Bev prepares lunch. It means opening tin cans. The Mackenthuns don't even drink their own milk.'

Malle first visited the Mackenthuns in 1979 as a part of a wider film about the farming community of Glencoe, and found a happy couple optimistic about the future. Delayed by other projects, Malle didn't finish *God's Country* until 1985, during the second term of the Reagan presidency, when globalisation was in full swing. He found a very different Glencoe. The prices of milk and grain had plummeted, forcing farmers to compete with each other, and with cheaper international imports. 'Glencoe farmers are having their best crop in two decades. They think it's a disaster.'

Back at the Mackenthun kitchen table, Jim and Bev again clasp their hands and say grace. They look to have aged much more than the six years since we last saw them. Bev has taken a job in town to alleviate financial pressure; they both look tired. In 1979 they hoped their boys would one day take over the farm; now they hope they'll go to college. 'It's sad to me,' says Jim, 'cos I like to farm. But I'm wondering if I'll still be here ten years from now.' What hasn't changed: the blinding whiteness of the supermarket bread on their plates.

I'd never much thought about all this until I found myself barging into restaurant kitchens, pushing open doors with a knee or shoulder, my arms heavy with boxes of figs I'd picked earlier that day, sometimes within the hour, the tops of each one glistening with white sap where I'd cut them free from the branch.

I was hardly in a position to criticise my Dad's functional relationship to food. The scrag-end of my twenties often saw me doing my groceries *behind* the supermarket. The thing about bin-diving is that it has a way of inverting the healthy food pyramid. What you eat isn't determined by nutritional advice, but by what you can reach from the top of the dumpster.

And so I lived in a series of share houses where we consumed more flavoured milk than olive oil, took bags of stale ham-and-cheese rolls to university for lunch, and offered guests dented cans of Fanta, presumably dropped by some zit-faced shelf-packer before being thrown out.

The peculiar nature of Canberra, a small city granted outsized importance by its status as the national capital, where you could take your leave from a power lunch and be on a farm before the bill arrived, meant I was suddenly linking the orchard to the table: my figs picked from the tree, weighed on the kitchen bench, then ferried from paddock to plate with the aid of my hand-me-down Holden Barina and cardboard boxes I'd scavenged from outside supermarkets (they'd originally been used to transport produce – Queensland bananas, New Zealand kiwifruit – from much farther away).

In a small way, I was closing a loop between production and consumption. I supplied five, then six, then ten restaurants – at this stage from the mother tree and the teenage fig trees my father had planted in the lower orchard – and rarely left a kitchen without tasting what was cooking. Most chefs didn't tinker with the figs much, slicing them in half and drizzling them in olive oil or honey or pairing them with soft cheese.

It became an end-of-harvest tradition to take my mother to lunch and dinner at restaurants serving Gollion figs. At Monster we ate them with buffalo mozzarella, olives and lardo; another autumn they featured in a duck salad. Bar Rochford served them with taleggio and slices of preserved lemon, garnished with miniature leaves of sorrel. Dad thought restaurants were a rip-off and was rarely convinced the food served there was better than what he could eat at home. He didn't drink alcohol but brought his own water to restaurants ('I can't *stand* the chlorinated stuff'), perhaps the most passion he showed for taste. Gollion's rainwater was delicious, and, without fluoride in it, perhaps the reason my teeth were so bad. The bin-diving years wouldn't have helped.

My figs weren't feeding the masses, and the people they did feed were a privileged minority. Until I had enough stock for the local farmers' market (itself the preserve of educated high-income earners), this would remain the case: the delicate nature of ripened figs means they couldn't travel much further anyway.

Soon chefs began asking what else they could buy. I brought them boxes of fig *leaves,* which I didn't even know were edible, and they made ice-cream from them, which didn't taste like figs at all but like coconut, rich and heady.

I filled the boot of my car with nectarines, peaches, apples, celeriac, rhubarb. One year Sean McConnell made bitty toffee apples out of our crab apples; another year he made sorbet out of our lemon verbena. What about our beef, Louis from Bar Rochford asked casually one day. Was that any good? I couldn't say. We occasionally sold lambs to market and had always eaten them, but we never ate our cattle; we always sold them. Hens too – we ate their eggs, but not their meat. Why not?

Dad was a pure economic rationalist when it came to cattle. With a bovine, he explained, the proposition was so big (an average carcass weight for adult cattle is 300 kilograms versus 20 kilograms for a lamb) that you need to know what you're doing. Employing a mobile butcher to slaughter a cow on site would cost a few hundred dollars at least. Then there was the meat.

'Do you have *any idea* how much meat comes off a steer?'

I didn't.

'You can only eat so much meat yourself. What are you going to do with the rest?'

Freeze it? Barter it with friends? Sell it at a farmers' market or direct to customers?

He shook his head. A look of dismay. I could tell he was imagining the bureaucracy required to bypass sale yards and slaughterhouses: regular hosings-down of on-site facilities; evenings hunched over packaged cuts of meat, applying stickers with marketing slogans; regular visits from safety inspectors, officials in white coats with clipboards.

'Then there's all the offal and waste products that aren't worth anything in sheep, but are valuable in cattle. You'd be losing all of that.'

*Nothing goes to waste*: the old peasant adage that is the foundation of the world's great cuisines. (In Italian, *cucina povera* means 'poor food', but now means 'authentic food', a repurposing that shows the efficiency of capitalism.) With no peasant tradition, Australia easily shed these conventions.

On one level, Dad was right: home butchering a steer is impractical and time-consuming. Peasant work is labour-intensive. But a pre-industrial Australian farm family would've had more children than a modern one, and, if they were lucky to live that long, more old people. More mouths to feed made killing a steer more sensible, but also meant there were more hands to help at harvest time (and with preserving, curing, salting), meaning time-pressed farmers didn't have to resort to tinned junk on their rushed lunch break.

For seventy years my father has eaten the same sandwich for lunch: ham, cheese, iceberg lettuce, tomato. It has never changed, from his Melbourne childhood to his Gollion adulthood, or from late summer (when tomatoes grow at Gollion) to mid-winter (when they don't).

There'd been a bull once, I remembered, that had 'broken down' in the paddock at Gollion some years ago, its shoulder fractured in a fight. Dad had euthanased it with the gun before it could be put on a truck to market. I was overseas at the time but remember Mum filling me in on the phone: rather than call a mobile butcher, a friend of a friend, an epicurean Frenchman named Claude, had done the deed, bringing his own knives and cleavers and leaving Mum and Dad with a freezer full of cuts, bagged and labelled in his mother tongue. They'd barbecued *bifteks* and made *boeuf* bourguignon, before the freezer was accidentally left unplugged one summer's day and defrosted, and they'd had to throw the rest of the meat away.

To farm your own cattle but not eat your own beef: it seemed a strange sort of disconnect, and one that would be absurd in other vocations. A winemaker who doesn't drink.

A preacher who doesn't read scripture. A chef who doesn't taste the soup bubbling on the stovetop, checking if it needs more seasoning. Loving the food a farm produced seemed like the logical flipside to loving the land itself. But my father was a retired free-market economist: being a specialist in a specialised economy was what he had once advocated for a living. It was entirely appropriate he would approach agriculture the same way.

There was something else he'd brought with him from the office to the paddock. So much of conventional farming is about intervention: meddling with the self-organising, self-correcting complexity of nature by growing vast annual monocultures, or reducing soil fertility to three chemical elements: N, P, K. But what is ostensibly done for efficiency inevitably results in more work: annual crops need to be replanted every year, and as monocultures they are left exposed to disease; synthetic fertiliser sounds good, but it can so disrupt the makeup of soil that it needs to be applied again and again to work.

Embracing the complexity and mystery of ecology, instead of attempting to simplify and control it, is the laissez-faire attitude to farming – the faith in a different kind of invisible hand to pick a winner. I had never understood him better.

# 8

Of all the things that can go wrong on an Australian farm – a locust plague eating your harvest; a combine harvester eating your hand; not enough rain; too much rain; a 'shy' bull; the prime minister showing up at the front gate for a photo-op during an election campaign – having a visitor sprain their ankle strikes me as one of the least concerning. But since I'd started working with my father, I'd realised that bushfires and cattle prices didn't worry him half as much as the public liability bogeyman.

It had happened to a friend of his, a farmer who had, in Dad's telling, graciously opened his gates to the public for an open day, only to be sued by some ungrateful klutz who had stumbled on a divot (or fallen down a rabbit/fox/wombat hole; the details, liable to change with every telling, weren't the point).

I believe the origin of his neurosis was the cautionary tale of *Liebeck v. McDonald's Restaurants* (aka 'the hot coffee lawsuit'), in which Stella Liebeck, a 79-year-old American *grandma* (that shocking detail was included in all the reporting), bought a cup of drive-thru coffee one day in 1994, spilled it on her lap, then sued McDonald's when she suffered third-degree burns. For a man who valued self-reliance and personal responsibility above all else, including and especially his own safety, this was the most egregious display of freeloading he'd yet encountered. America, whose rugged individualism you'd think would appeal to my father, had in his mind 'gone mad'; a land where the ability of your attorney held more sway in determining your lot in life than any bullshit adherence to the American Dream. His friend being forced to pay damages for a sprained ankle was proof that the scourge of litigation had made its way across the Pacific.

And so, he didn't allow visitors to ride their horses, or their bicycles, or our motorbikes. Breaking his own bones falling from a hayshed roof didn't concern him as much as the vision of being served a summons should he be foolish enough to employ someone to help fix, and fall off, the roof with him.

When I was twenty-one, I rolled my ute one evening on a dirt road not far from Gollion. I'd been speeding, deliberately so, taking a corner of built-up gravel too fast so I could momentarily 'get sideways', in the parlance, as I turned. This time I misjudged the corner, got very sideways, overcorrected, and ended up crawling out the passenger side window, my ute upside down, stopped from rolling further by the roadside fence.

I learned a few things that night. For starters, I've never gotten sideways again. Then there was the famous Australian value of mateship, displayed in the first vehicle to arrive on the scene, two blokes in another ute, who, finding me crouching by a wrecked vehicle, clearly in shock and not having yet called my parents, congratulated me on my 'party trick', before the driver floored the accelerator, their laughter ringing in my ears.

When Dad finally showed up with his own ute and a towrope, he was frantic. He hurriedly attached the rope between his ute and mine, was able to pull my ute back upright and back onto the road, and ordered me into my driver seat, now crunchy with windscreen glass; I was to put it in neutral and steer as he towed me home. 'If the police show up,' he panted as he put on his hazard lights, 'tell them you swerved to avoid a kangaroo. If they're paying attention, they'll look at your tyre marks and use them to book you for speeding!' It was a weekday night on a quiet country road. The police weren't showing up.

It was a lesson in the extent of Dad's fear that everyone would take advantage of you: private litigators, insurance companies and the state. I already knew that for Dad, a piece of rusting wire lying around the farm represented a tetanus lawsuit; the air compressor, which had a leak in the hose that caused it to loudly turn itself on every ten minutes when plugged in, was *not* to be replaced ('a good burglar alarm, that'). Now, I also knew that crashing your car came not just with the risk of self-inflicted injury, but the threat of an investigation as well.

He loved living on a farm, he often said, because there you're 'free to do whatever you want'. In America this is shorthand for 'shooting anything and anyone that steps onto your property,' but it wasn't like that here. There were none of those TRESSPASSES [sic] WILL BE PROSECUTED signs you see tacked to trees in the bush; he wasn't the aggressor in this fight. Freedom for my father meant being able to urinate on his lemon tree in broad daylight, or to drive the unregistered East German Trabant he'd had shipped over from Bremerhaven at great expense while less imaginative souvenir-hunters were chipping off bits of the Berlin Wall. It wasn't libertarianism – he believed in a strong welfare state for those less fortunate than himself. He just wanted to be left alone.

Private property is a central tenet of this concept of liberty. Freedom – from prying eyes, from a neighbour's disease-ridden livestock – must be protected by a clear demarcation between what is yours and what is not. Gollion's boundary fences were built with an extra line of barbed wire compared to the internal fences, and with the steelies spaced closer together. It was the same for other boundary fences in the district – my ute would likely have kept rolling if the roadside fence hadn't been so strong.

Self-reliance was another pillar of farmyard freedom, and Dad was proficient in an impressive range of fields. He could weld, join, build, shear, rig up electricity and do his own plumbing. Irrepressible, what he couldn't do he'd 'have a go' at anyway. (Watching football or cricket, at the ground or from the couch, he liked to yell, at any sign of athletic hesitance, 'HAVEAGOYAMUG!') He disdained tradesmen, whom he considered arrogant and predatory, overcharging for a service that 'any old mug can teach themselves'. For building jobs that required 'a touch of class' (anything inside the house that Mum would use) he employed Bob the Builder, a raffish old boozer who had done his apprenticeship in the 1950s. In an inversion of the employer–employee relationship, Dad paid Bob but also laboured for him. Bob taught Dad the way Dad taught me: by example.

My father was aware that this was an unusual way to behave as a seniors-card-carrying member of an affluent society in the early twenty-first century. He called himself a 'fifties boy'; a man whose mid-century childhood involved 'mucking about' with crystal radios and Holden engines. He lamented the demise of the do-it-yourselfer, a trend he blamed on government overreach (the law requiring a ute-load to be covered by more than a few frayed ropes was making his proficiency with knots obsolete), and the self-interest of companies that made it impossible for their products to be repaired at no cost ('I *hate* modern cars – everything's done by computers. It means you can find the problem but you need a specialist to fix it').

The downside of doing things yourself is that it does nothing for your social skills. There are no public roads dissecting Gollion; my father is not what one calls 'street-smart'. There is a scene from my childhood that comes to mind whenever I hear this expression. We are in Melbourne, on a family holiday in the mid-1990s, in traffic. Dad is behind the wheel and is arguing with Mum about directions. He is stressed and driving erratically. He changes lanes without indicating: a toot and a flipped bird comes from an aggrieved driver behind him. Dad pulls over, confused and now even more stressed. He clicks his seatbelt off and steps into the traffic, running his hands over the top of our family van. 'There must be something on the roof,' he explains. 'That bloke kept pointing at our roof with his finger.'

One day in early 2017 we were priming the siphon in order to irrigate the orchards. *Priming* in this context means flushing air out of the pipe before use, achieved by first turning off all the taps near the house to close the line, then turning on the pump connected to another, *lower* dam, in order to push air *up* the hill to the siphon dam. There, it must be removed from the line by opening a plastic cap, which is then screwed back on after water starts rushing out the pipe behind the air. You then have a window of about ten minutes during which the pump from the lower dam is pushing water up a line that you have just sealed with a cap – a pressure build-up that will eventually blow up the pump if its switch isn't turned off by rushing back down the hill on foot or a motorbike. The siphon is then ready for use.

That day, we were going through the steps again: despite doing it several times before, and despite my keeping a mud-map (drawn by Dad) outlining the process in my wallet (like a cartographical map, it had to be folded in a certain order, which I could never get right, making my wallet hard to close), I couldn't remember the steps, or if I did I remembered them in the wrong order. I once overfilled a tank for several hours, flooding the adjoining paddock. Another time I waited too long to remove the plastic cap and release the pressurised air; by the time I did, it blew out of my hands with the force of a high-school science experiment. Thankfully, I hadn't yet blown up the pump.

Dad didn't mind giving me a refresher course: much like a golden retriever, he couldn't pass a body of water without getting wet. Hydraulic engineer was up there with orchardist and Aldi tool tester as his dream career if he 'had his time again'. ('I just *love* working with water – the smell, the taste, the way it moves through the landscape . . .') While we waited for the bubbles to disappear from the siphon pipe – a sure sign that the oxygen had been flushed out of it and we could re-attach the cap and proceed to turn off the pump – he continued today's lesson.

'You don't get too many free lunches in farming, but gravity and sunlight are two.'

The free lunch: another concept that married my father's vocations of farming and economics.

I screwed the cap on as tightly as I could. 'You love a free lunch, don't you?'

He clicked his tongue in agreement.

Another freebie, he said, was a dam that hadn't been made properly watertight. This dam, despite its leaky wall and despite it being our chief irrigation reservoir, was largely full. But as we walked down its eastern slope, I saw that a lush strip of grass was seeping *through* the wall, testifying to the dam builder's fortuitous incompetence.

Dad stopped at a tap fifty metres below the dam; it was one of those little metal wheels you see submariners turning in recruitment ads for the navy.

'If the siphon isn't priming for some reason and you've tried everything else,' said Dad, 'sometimes *someone* might have turned off this tap: a stranger up to no good or some naughty kids visiting, like the Pearce boys.'

'The *Pearce* boys?' The Pearce boys were the children of family friends. I hadn't seen them in maybe twenty years. 'They'd be in their forties now!'

He laughed as he checked that the tap was tight, but then he doubled down. 'Well, you know what I mean – they used to do stuff like that.'

And then I told him I'd invited some strangers onto his farm – not a few strangers, with a few ankles that could be sprained, but *thirty* strangers totalling *sixty* strainable ankles, many of them strapped into flimsy high heels.

The occasion was a long-table lunch and the promise was, in the words of the marketing blurb, 'Paddock to plate, on plates in a paddock'. My friend Harry's father was more cynical: 'Is there a *reason* for this event, or is it simply a brazen money-making venture?'

It was both. Or, more accurately, the former enabled the latter. A showcase of local, seasonal food, cooked and eaten where much of it was grown and harvested, would (I hoped) beget a money-making venture. Where most farmers in developed countries face pressure to keep prices rock-bottom, this event would remove the many stages that come between the farming of food and the eating of it. It would avoid the wrangling with middlemen and the out-of-season ingredients flown in from another hemisphere; it would minimise fossil-fuel consumption; and any modest profit would be retained by me and my event colleagues, members of the local community who would produce, cook and serve the meal.

The idea came from my housemate 'Little Soph' and her girlfriend, Lean. Soph was a waiter at Bar Rochford; Lean was a photographer and event stylist. Louis Couttoupes, who worked with Little Soph, agreed to cook the meal. His partner Iwona would be sous-chef. By now Louis had a following as a rising star of Australian kitchens, a walk-in who had talked his way, with zero professional experience, into an unpaid internship at the legendary Parisian bistro Au Passage. Returning to Australia, he was hired at the newly opened Bar Rochford, initially doing dinner prep and sometimes even security. When he gained control of the kitchen, he served European classics wryly innovated: salt-and-vinegar potato galettes dusted with dehydrated bush tomato; tomatoes (this time the garden variety) and stracciatella with a squirt of Gollion fig-leaf oil, a dab of mango vinaigrette and a topping of bee pollen; spatchcock with tarragon butter roasted in hay ('a typical farmhouse roast chicken, with extra farmhouse').

We were confident the cashed-up young urban professional crowd who ate his cooking at Bar Rochford would trek out to Gollion for more. On coming home from farm days, I'd been speaking to Little Soph about my desire to extend my reach into the food system; to better control what happened to our fruit and livestock once it left the farm gate, rather than watching it disappear into the maw of the market.

We picked a date, a Sunday in mid-March: reliably warm but not blazing weather to set up a table in the top orchard and, by my estimate, the peak of the fig season. After factoring in food and drink, a liquor licence, transport (we would hire a minibus to ferry guests to and from what would be an all-you-can-drink affair), we decided on thirty guests to make it profitable while keeping it intimate. Louis would forage what he could to cut costs (he had a slew of secret roadside spots), with additional vegetables provided by local market gardeners Brightside Produce. The wine would come from our friend Sam Leyshon, a one-time Parliament House security guard and rock-band frontman who, like Louis, had spent time in France learning from masters. He now made experimental cool climate wines on his family's vineyard, Mallaluka. Gollion would provide two lambs, seven kilograms of figs, orchard ambience and a pre-meal tour by the event host, 'Farmer Sam'. Tickets went up on Lean's blog in February: $150 for lunch and transport; $125 for lunch only. They sold out in ten minutes.

Correctly anticipating Dad's response, we had factored public liability insurance into the budget. But he had other arrows in his quiver. How would we legally sell meat we'd slaughtered to members of the public? Last time he looked, none of us were accredited butchers, a necessity to meet food and safety standards for restaurants, pop-up or otherwise. Even if we could legally sell the meat, how would we cook it? Louis wanted to cook the lamb, its four legs bound in herbs and hanging from a tripod, over hot coals, but the wet winter of 2016 had given way to a dry spring and summer, creating a heavy fuel-load in the Australian bush. Did we even have enough figs? The winter rain may have dried up, but it had delayed the start of the crop.

We couldn't speed up the ripening of the figs, but we could address his other concerns. I called the Yass Valley Council and confirmed that it was illegal to sell meat to members of the public without the adequate butchery accreditation. They suggested a mobile butcher, who would slaughter the lambs on site; a portable cool-room could be hired to hang the carcasses. The fire captain of the Sutton Bushfire Brigade was equally intent on making life hard: throughout the month of March, he said, no open fires would be permitted in the district.

We were at a crossroads. The choice: spend more money and follow official advice, or get creative. If the lamb, I discovered, never left Gollion – if it were killed by me and Dad, hung in the shed, butchered and then cooked by Louis – it *technically* wasn't entering the marketplace, so was exempt from the rules. And if the price of the ticket covered the event and not specifically the meat, we *technically* weren't selling it. (If anyone asked, we'd say the lamb was a bonus – the gift that came with our Happy Meal.)

As for the fire, a pizza oven Dad had built in the early 2000s (which had itself once caught fire, on account of Dad using wool to insulate the corrugated-iron roof – without first removing the sheep shit from the wool) was enclosed, but too small to fit two lambs inside. I suggested creating the coals by burning wood inside the pizza oven and moving them to a pit elsewhere to cook the lamb, until Dad pointed out that all our wheelbarrows were either made of plastic or pockmarked with coal-sized holes.

Poking around Fernleigh, Louis and I found a better idea: an old fireplace flue, which we took back to Gollion and perched above a wheel-less car trailer, the whole thing lashed together with four steelies and some wire. The trailer would be our firepit; the chimney our legal defence that we weren't cooking over an open flame.

Dad had been appeased. What mattered wasn't that we didn't start a fire or make people ill, but that we wouldn't be prosecuted for doing so. Slightly dodgy but technically legal: my father couldn't argue with that, and nor did he try. I think he was even impressed.

The figs, meanwhile, were still green. Louis had been busy foraging from the roadside – plums, fennel pollen – but

Gollion's black Genoas were a highlight of the menu. The idea was that our guests would be eating fruit that had grown an arm's length away; the table would run across the top orchard, with the fig tree's lower branches brushing against the seats.

The rest of the menu, with the exception of the flour, salt, pepper and butter, was produced by local farms. What defines *local* is notoriously tricky. 'Australian-grown' doesn't mean much when the produce in question might have come from a climate and several time zones away. I checked the website of my nearest farmers' market (thirteen kilometres away) but couldn't find a definition. Perhaps I was overestimating the public's food literacy: questions on the FAQ section included: 'What defines a producer?' and 'How can I pay for my goods at the market?' It was only under 'Market rules' that I found this line: *Food from the market comes direct from the producer, with the majority of produce coming from less than 300 kilometres.* Three hundred kilometres: the distance from Brussels to Paris, from New York to Boston.

If 'local' denotes familiarity – the same climate, the same terroir – we set ourselves a limit of an hour's drive from Gollion. You'd have to break the speed limit on some corners to make that true of Brightside Produce, the chemical-free market garden that provided the vegetables. But the hay in which Brightside's purple carrots would be cooked offset that concern: it came from Fernleigh's hayshed, where it had been maturing for over ten years.

The rest of the menu began to take shape. On the Thursday before the event, Dad and I killed two lambs, repeating the process I had first learned in the spring of 2014, and a few times in the years since (each time there was some new nugget of knowledge to impart; on this occasion Dad told me to cut open the liver to check for fluke: 'I've heard of farmers cutting open a liver and a cloud of moths flies out at them . . .').

Lean and Little Soph scouted the property for decorations. From Fernleigh they returned with hardwood timber and antique toilet doors, an old horse watering trough, two sun-bleached cow skulls and some vintage drawers made from Shell motor-oil tins.

With Dad's help, the dining table Lean used for functions was erected in the top orchard, running perpendicular to the slope between two rows of trees, its legs chocked with blocks of timber to make it level. Umbrellas were hung from apple branches and vintage leather satchels arranged underneath. An old butchery table was wiped clean of blood and lugged inside the orchard for the wineglasses. The Shell drawers sat beside them, and inside was placed a native floristry arrangement that didn't look so different from a bundle of kindling. It was all very aspirational: the kind of bucolic fairy-tale where brides are more likely to gambol than lambs; where vows are exchanged but not a lot of animal husbandry actually goes on.

Mum enjoyed free lunches as much as Dad did, and I bought her a seat at the table. But my father thought it all looked 'a bit much' for him. He would spend the afternoon watching movies in Canberra. He hated Gérard Depardieu, but the French Film Festival was on.

Unlike the minibus Mrs O'Toole drove us to primary school in, this one stopped at Gollion's front door. The guests alighted with the nervous energy of a group of acquaintances randomly thrown together and forced to socialise. It turned out several of them were known to us, if not to each other: friends of friends, or 'hospo' colleagues. The ticket hadn't specified a dress code, but the guests looked like they were off to someone's wedding.

I had two jobs for the day, chief among them hot-coal liaison officer. The flue was successfully drawing smoke from the fire underneath it, and the trailer, surrounded by sheets of corrugated iron, gave added protection from errant sparks. With its numberplate still attached – B-25321 – this was the ultimate paddock basher: an unregistered farm vehicle left to rust and lose its wheels, and now set on fire.

With a long-handled shovel I scraped coals from under the flue to the front of the trailer, above which cooked four lamb legs. S-shaped hooks pierced their Achilles tendons and suspended them from a horizontal metal pole lashed to two upright steelies. Louis had trussed the legs with butcher's twine and stuffed each one with whole fennel plants and bay leaves. They sizzled as they cooked, fat spitting on the coals underneath.

My other job was farmyard prop. The morning of the lunch, Lean had asked me to move the sheep into the paddock above the top orchard to create more photo opportunities; they had no grass to eat there, so I refused. I was, however, happy to play dress-ups. I wore chinos and work boots (my scruffiest pair); my linen shirt had been selected on account of its tears, and I made sure to undo the top few buttons, exposing a budding red farmer's décolletage. My Akubra, too, was deteriorating beautifully, now covered with spots of mould and no longer flat-brimmed but curved by a rainstorm then hardened in the sun, the way waffle cones are shaped while the batter is soft.

Pét nats (I had to google them that morning) were served off one of the toilet doors in Ram Paddock, bottles on ice in one of Fernleigh's old washing coppers. As guests sipped and admired the view to the valley below (any farm junk we could move had been hidden in a shed, the agricultural equivalent of sweeping it under the carpet), I welcomed them to Gollion. I gave a brief introduction to what we grew, the history of the property and how we farmed, emphasising that the lunch showcased organic produce, cooked by a local chef and eaten where it was grown. I walked them through the bottom orchard, explaining how and why I was growing my young figs (an unimpressive sight: the orchard net was not yet complete and the figs were only knee-high) and then through the top orchard, surprising myself at being able to identify every tree, before ending the 'farm tour' at the dining table, where thirty places were set for lunch. Our guests took their seats between the old fig tree, two pear trees and a cherry.

Back at the trailer, Louis prodded the lamb legs and hurriedly plated up the entrées.

***First***

*Fig, fromage frais, honey*

*Lamb tartare, kocho, wasabi root*

*Smoked duck, thyme, viola*

***Second***

*Lamb cutlet, wild plum, pistachio, rose*

***Third***

*Tomato water, lovage, nasturtium*

***Fourth***

*Poacher's lamb, wild fennel, figs*

*Purple carrot, labneh, black barley*

*Beets, black walnuts, burnt orange*

***Dessert***

*Hay panna cotta, milk crumb, pear*

It was unlike anything that had ever been cooked at Gollion, from local ingredients or otherwise. My mother assured me the meal was as good as it sounded, although she had been seated next to a 'painful' young diplomat who just wanted to talk about himself and get drunk. Was this the only way to make money from farming in Australia? Cater to the aspirational, managerial class with Instagram-worthy meals and settings? It wasn't exactly feeding the masses. It was, at least, sustainable in other ways. The UN says one third of all food produced for humans goes to waste, but our event didn't waste a thing.

And in the end, we had just the right number of figs. The wet winter had delayed their ripening, but they had proved perfect, picked the day before the event.

That the event had gone spectacularly well wasn't a surprise to me: we had worked hard, Louis had created a thoughtful menu, and he'd pulled it off with the help of Iwona, Lean and Little Soph. What did surprise me was the ease with which I'd assumed my 'Farmer Sam' persona. I'd had no problem answering the guests' questions about property size, rainfall, the allowable insecticides in an organic orchard (copper oxide and lime sulphate). I'd explained that figs have two crops, but the first, or *breba,* crop – mealy and flavourless – we pulled off the branches to encourage the tree to put its energy into the superior second crop. When someone asked me what kind of soil we had, I mumbled the only answer Dad had offered me when I'd asked the same question: 'Crap'.

I had convinced the guests, but I knew that was a low bar. I might have been a farmer in the eyes of a drunken audience of gourmands, but what about a paddock full of farmers? I needed to go to school.

PART III

# Grazier

The cowman who cleans his range of wolves does not realize that he is taking over the wolf's job of trimming the herd to fit the range. He has not learned to think like a mountain. Hence we have dustbowls, and rivers washing the future into the sea.

**Aldo Leopold,** ***A Sand County Almanac,*** **1949**

# 9

I went to grazing school.

It was the beginning of 2017, a hot, dry January (the roadside fire danger signs oscillated between SEVERE and EXTREME), and as I entered the third year of my 'apprenticeship', I was ready to drop the quotation marks. What we'd started together in 2014, my father and I, was a lark born of necessity. I, able-bodied but useless, would help him, knowledgeable but incapacitated, to keep the farm running while he convalesced. But he'd recovered years ago, and I'd stayed on. Dad hadn't said so, but what we were doing had *become* an apprenticeship of the old-fashioned kind, one with a master and a protégé. When it would end I didn't know, but I would know it had ended because then I would be a farmer. But as I'd learned from labouring on building sites alongside proto-carpenters and electricians during my university holidays years earlier, before an apprentice can graduate they must first obtain the dreaded 'ticket': the theory component of their training, inside a classroom.

It was my mother who suggested the course. She had run farming workshops for many years (land management planning; weed identification), and was always emailing me farming tidbits: links to 'field days' and 'seed swaps', working bees and workshops. Although Dad still occasionally grumbled about how much time I was spending with him instead of adjusting my cufflinks in an office cubicle somewhere, Mum seemed to encourage it. Gollion was her farm as much as his, and I took these emails as reminders that if I wanted to succeed them as its custodian, she would support me.

Not that this was ever made clear. We never raised my taking over the farm; I never said I wanted to, and they never asked. I can't remember a particular moment when Dad articulated that he was training me to succeed him, but I remember realising that it had become tacitly understood. It was during that summer of 2016/17, and we were discussing the large hill, previously part of Westmead Park, that looms above Gollion's chook shed. The old windmill that had once screeched at its base had toppled into the grass when I was a teenager. There was a row of poplars beside it, planted by Jack Whyte in the 1960s and now thirty metres tall. Since Jack's death, the block had passed through several owners and had recently been bought by a successful local businessman, whose son was building a house on its summit. The hill was so steep that sheep grazing its upper slope looked from Gollion to be floating above those further down, figurines on different levels of a shadow box. 'More money than sense', another neighbour had said to Dad about the son's plans for the house.

My father loved that hill – its Kellogg's Corn Flakes autumn foliage, its easy-to-defend vantage ('if this was Europe there'd be a castle on it') – and he often lamented not buying it when it had come onto the market a few years back, an acquisition that would have pushed Gollion past 800 acres. 'Oh well,' he said, handing me the pliers so I could attach a patch of chicken wire to the gate of my new fig orchard. 'He'll probably buy you out one day, anyway.' It was around this time I wondered to my sister Lucy whether Dad might begin paying me for my farm work. 'You're getting A FARM,' she replied. 'That's your payment!' For now, that was the closest we came to discussing who'd get to take over.

Nor was there a moment I knew I would farm Gollion when Dad eventually 'retired' (unlike my apprenticeship, his eventual, hypothetical retirement still seemed worthy of quotation marks, given his relentless work ethic). I can, however, pinpoint a moment I realised that maybe I *could* do this, which came before I realised I *wanted* to. It came during the winter of 2015. My parents were overseas, and I was staying at Gollion for the fortnight of their trip. Dad had written a complex set of instructions and left it on the kitchen bench, including when and where to move the cattle according to their rotation.

A few days after they left, I moved the cattle according to the instructions – but when I checked on the three bulls in a paddock near the house, I found that one was missing. Bulls fight; the vanquished sulk. I spent a whole sodden Sunday looking for it, phoning up neighbours, walking the boundary fence, slowly driving through the district, trespassing into other properties for a closer look at their bulls, the ear-tag ID of the escapee written on my hand in running blue ink.

Towards nightfall I was scanning a copse of acacias just outside our western boundary and there he was. Uninjured, alone, not cowering but standing tall, he looked at me looking at him. I cut through the wire fence, mustered him back onto Gollion, and then repaired the fence the way I had been taught. Riding the motorbike back to the house, I felt euphoric.

But my farming ignorance remained profound; Dad's knowledge was vast, but he alone could not teach me the skills I'd need to keep Gollion flourishing.

I enrolled in a course in holistic management, run by TAFE New South Wales and taught in two-day blocks over a period of eight months. *Holism* is derived from the Greek *holos* ('complete', 'entire') and posits that the world comprises a series of systems or *wholes*, autonomous from but dependent on each other, and that changes in one system have implications for others. It shares elements of James Lovelock's Gaia hypothesis, which holds that Earth is a living organism comprised of trillions of smaller organisms (a microbe, a whale), whose individual health determines the collective capacity of the planet for self-regulation and resilience. While the conventional farmer compartmentalises agriculture from nature, the holistic farmer believes the health of both is determined by the ecosystem they share.

Hardcore proponents of holistic management apply it to every decision they make (at the start of the course, we were each given a wallet-sized 'decision-making tool' for reference). When deciding whether to get out of bed, they first consider the impact sleeping-in would have on the various wholes of their life – family, work, finances, wellbeing. Applied to grazing, holistic management focuses on the healthy function of four ecosystem processes:

1. The solar cycle (the process by which sunlight is converted into energy by green photosynthesising

plants, feeding the plants, the animals that feed on those plants, and the animals that feed on those animals)

2. The water cycle (by which moisture is stored in and released by the landscape)
3. The mineral cycle (by which nutrients, minerals and chemical elements pass through the soil via plants, animals and microorganisms)
4. Community dynamics (by which all beings, including humans, interact with each other and with the three other cycles, upon whose functionality and interconnectedness they depend)

The course was taught by Brian, a jovial cattle and free-range egg farmer with a habit of sticking his thumbs in his belt loops and rocking back on his heels while he waited for us to answer his questions. It was the first time TAFE had offered holistic management in its curriculum, and I wondered if this was a sign of regenerative agriculture's growing stature.

Our classroom was a demountable building in Mongarlowe, a village ninety minutes' drive east of Gollion. The village was known locally for its resident platypus, and for its weekly men's group (they called themselves the trolls, because they discussed their masculinity problems under a bridge).

I had expected my classmates to be ruddy-cheeked, middle-aged cattlemen; instead, they were mostly, like me, tertiary-educated thirtysomethings who wanted to grow their own food. We were writers, a ceramicist and a filmmaker; a market gardener with a background in conservation; the manager of a local farmers' market and her partner, who fed his chooks on maggots foraged from road-kill kangaroos.

Holistic management aims to provide a framework for making decisions that will enable the four ecosystem processes to function at their best: leafy green plants photosynthesising for as much of the year as possible, over as much land as possible (the solar cycle); soft, aerated soil, alive with bugs and fungi and covered with plants, soaking up rainfall like a sponge so that it does not evaporate or run off the ground (the water cycle); nutrients brought up to the surface by deep-rooted plants and passed on to grazing animals, who return it to the soil in their dung and to predators in their meat (the mineral cycle – you are what you eat *eats*). The fourth landscape function, community dynamics, is like a census of the ecosystem, which ideally includes a complex bunch of plants and animals, diverse in size, age and maturity, interacting and often competing with one another in ways that benefit the other three functions.

The idea is that nature provides land managers with all the tools to achieve a healthy, well-functioning landscape; holistic management is about knowing *when* and *how* to apply them. To let nature help itself, we need to give it a leg up.

Brian told us that in parts of the world that are damp for most of the year – the Amazon, the Everglades, England, Japan – the interaction of the four ecosystem functions is relatively straightforward. In a tropical rainforest, where all available space is taken up with greenery, moisture comes in the form of consistent rainfall, which drums on the leaves, soaks into the soil and transpires back into the atmosphere via the trees. When a tree dies and falls to the forest floor, the humidity is such that it will have decomposed within months, bugs and fungi turning wood into mulch and then humus, the beginnings of new soil. These environments experience little seasonable variability, with humid conditions that support insects and invertebrates above, on and below the ground (the biology in the Amazon, said Brian, is 'on the job 24/7'). Holistic managers call an environment like this 'non-brittle'.

Conversely, an example of a brittle or 'brittle-tending' environment is one where all or part of the year is *not* humid. Think the Great Plains of America in winter, the African savannah during its dry season, or the Gobi Desert all year round. During these periods, the activity of photosynthesising slows or stops, and with it the solar and water cycles. The small mammals, insects, invertebrates and soil microbes that cycle nutrients above and below ground when the climate allows – these either die, pupate or become dormant. In this breach, argue holistic managers, nature has evolved a clever solution. Biology might be on holiday outside, on the frozen steppe or the sun-bleached veldt, but *inside* the stomachs of the few animals that remain, it is working furiously. 'Mobile recycling units', Brian called them. My textbook called them ruminants.

Like our guts, the first stomach of a ruminant (their rumen) contains tailored microbes that assist them to digest their diet – in a cow's case, the tough, fibrous part of plants called cellulose. Humans can't digest cellulose, but ruminants can because microorganisms called methanogens ferment it inside the rumen.

Until the 1960s, the conventional wisdom of pastoralism held that overgrazing – whereby an animal eats a plant faster than the plant can regrow – was caused by having too many mouths to feed. It was Allan Savory, a young Rhodesian ecologist, who popularised the idea that what mattered was *time*: how long an animal spent munching on a particular plant; and how much time elapsed before they returned to that same plant. If they grazed on the same plant for too long, or returned to it too quickly, before it had a chance to regrow its leaves, the plant would have to rely on energy stored its roots. If this happens repeatedly, giving the plant no chance to access new energy through photosynthesis, it would eventually use up its stored energy, die, and leave bare ground – changing the micro, and eventually the macro, climate.

Savory reached this conclusion following what he came to call 'the saddest and greatest blunder' of his life. Tasked with setting aside land for future national parks, he assumed that the degradation of the de-peopled, destocked land was due to overgrazing by elephants. He crunched the numbers and determined that the land was overstocked. In the following years, 40,000 elephants were shot. But the land continued to degrade. Savory emigrated to the United States and was shocked to discover the same thing occurring on land that had been free of grazing livestock for over seventy years. Contrary to prevailing ecological logic, it was becoming more denuded with time.

Savory also hypothesised that it was possible to 'rest' pastures between grazings for *too* long. This is where his idea of 'brittle' environments comes in. In non-brittle environments, bare ground doesn't stay bare for long before something green covers it up quickly. The process by which organic matter is recycled happens quickly and naturally through consistent, year-round humidity. But in landscapes where desertification is occurring (semi-arid North and South America, Asia, Africa, Australia), moisture is only guaranteed for a few months of the year, if at all. In his 1988 book *Holistic Resource Management*, Savory writes that, over time, 'a lack of adequate physical disturbance' causes the ecological community to become 'simpler, less diversified, and less stable'. Without ruminants recycling them, brittle grasslands become rank and decay, eventually returning their carbon back to the atmosphere in a process called oxidisation. You can tell an oxidising plant, said Brian, because it is grey.

When humans first domesticated herbivores, they did so in grasslands with distinct variations in seasonal rainfall – the Fertile Crescent, the steppes of Eurasia. The soil and vegetation in these ecosystems had developed through the interaction of large herds of grazing animals and the pack-hunting predators that ate them. The main defence of a wild grazing animal is its herd: the bigger the herd, the safer the individual. Bunched together by pack-hunting animals, the herds spread dung and urine all over the pastures they grazed and were kept moving by the predators and the need for fresh grass. This constant movement, argued Savory, prevented overgrazing, while the trampling of what they hadn't eaten into the soil ensured a layer of mulch or groundcover. Bare ground gets hotter and loses moisture faster than mulched ground. It was, I remembered my father saying, my enemy.

When the first pastoralists tamed the herbivores that would, with breeding, become the antecedents of today's sheep, goats and cattle, they removed the threat of predation – and with it, the impetus for movement. Nomadism and transhumance, with its emphasis on migratory pastures, retained a semblance of the old rhythms of nature, but stationary farmers replaced the roaming wolves and lions with fences, limiting the herbivores' grazing range. This is known as 'set-stocking'.

In 'brittle' environments, cellulose can be broken down and returned to the soil. In such environments, Savory asserts, ruminants chewing their cud are playing the role played by the steamy floor of a tropical forest in a non-brittle one.

The implications are startling and counterintuitive: desertification is not a natural process, but is caused by animal husbandry, and the only solution is achieved through better – more *natural* – animal husbandry.

In the decades since, Savory has applied holistic management to his 20,000-acre Zimbabwean ranch, Dimbangombe, where his 500 or so cattle aren't corralled into fenced-off paddocks, but graze intensively inside a two-metre-high portable canvas fence, which is supported by trees and encloses an area of approximately an acre, guarded by stockmen and dogs to ward off lions. After no more than three days (in the dry season) and sometimes only one (in the wet season), the canvas is rolled up and moved to fresh pasture that has been rested for at least two months. The result is increased groundcover, biodiversity, more carbon being returned to the soil via mulching, and better-functioning solar, water and soil cycles.

Savory is a pariah to traditional rangeland academics, who maintain that desertification is caused by changing rainfall. His claim that, by applying holistic management to just half of the world's grasslands, we could store enough carbon in the soil to reduce atmospheric carbon to pre-industrial levels, has divided opinion and attracted scorn. But his principles are now applied to 15 million hectares on five continents, and the before-and-after photos Brian showed us in class suggested there was merit to the counterintuitive logic that *intense* grazing is better than no grazing. It was one of the 'paradigm shifts' Brian encouraged us to take in, rocking back on his heels in silence as he watched us process what he said. Overgrazing wasn't caused by *mouths* but by *time*; if livestock are set-stocked, farm fences are effectively prisons for animals and result in overgrazing; putting in *more* fences, especially portable ones, gives the pastures outside them a chance to rest and so results in more grass. *More* cattle, not fewer, was what the planet's grasslands needed. (I thought: *Really? You haven't even mentioned methane yet.*)

In the past few decades, said Brian, a growing number of holistic graziers have begun calling themselves 'grass farmers' – they may sell wool or meat, but their capital is soil and pasture. More than anything, the health of the farm depends on the health of the soil.

Grass farming is management-intensive: the rate of growth and the pace at which cattle or sheep are moved around changes according to the season. Generally, animals are moved more slowly in winter and summer, when the grass is growing slowly, and fastest in spring and autumn, when it experiences a 'flush' of growth. Overgrazing is more likely to occur in these shoulder seasons, if a plant that has been shorn of its leaves is chomped again before it has fully recovered. There are myriad variables to determine the rate at which a paddock recovers from grazing: season, sunlight, rainfall, as well as the age of a cow and whether it is pregnant or has a calf at foot (a lactating cow will eat up to twice as much as a dry one).

To keep track of and plan for this complexity, Allan Savory has created a 'grazing plan and control chart', based on British military training techniques in which strategists are taught to take into account ever-changing battlefield situations. Savory's resource allows graziers to plot their livestock movements on a daily basis, and to factor in multiple variables (when a bull will be joined with the cows; when a dam will be empty) over months or even years. Brian encouraged us to use these charts, and I pinned one on Gollion's kitchen wall beside the rainfall chart. Dad encouraged me to fill it in but said he had never before felt the need to, because he kept track of the cattle's movements in his head.

Getting animals to the right place at the right time for the right duration can result in more grass (because intensive grazing prompts new growth) and builds new soil (because leaf litter and manure carpet the ground, and bacteria, fungi and earthworms break these down into humus).

Done well, said Brian, grass farming is a process of *monitoring*: employing observation and local knowledge to make minor adjustments. A cattle farmer manages a hundred or a thousand head; a grass farmer manages hundreds of thousands, if not millions, of *seed* heads.

I wondered if my father knew he was a grass farmer.

# 10

Brian gave us homework, plenty of it, but my parents were happier to help this time around than when I used to bring home algebra in my schoolbag. I spent hours traipsing through Gollion's paddocks, the sweat slippery-dipping down my nose and blotting my worksheets. We were asked to find and photograph examples of the four ecosystem processes in various states of (in) action: overgrazed plants; plants recovered from grazing; signs of an ineffective water cycle; bare, compacted soil.

Mum had studied botany at university and had run many 'weed ID' workshops; she helped me to tell the xerophytic plants from the hydrophytic, and the cool season grasses from those that grew in summer. Dad had a mental map of the farm and pointed me towards signs of regeneration: patches of bare ground that had covered over with grass in the years since he'd started to holistically graze his paddocks; watersheds that now stayed spongy for weeks after rain instead of days, as was the case when he and Mum bought the place.

Everything Brian asked us to find was where it was, in the state it was, because of human management. Overgrazing happened if Dad kept a mob of cattle in one paddock for too long or let them return to it too quickly. Annual grasses favoured ground that had been turned bare by animals; perennial species were a sign of a better adherence to 'the law of the second bite' – the most sacred decree of holistic grazing, in which herbivores must never be allowed to eat a plant that has not yet recovered from being recently grazed.

In the growing season, Brian told us (at Gollion this corresponds to spring and autumn), it is hard to overgraze a plant if the animals grazing it are moved to a new paddock within three days. This is because even with ample rain and sunlight, a temperate plant will not regrow an eaten leaf immediately, and in the meantime can survive on energy stored in its roots. 'Use your lawnmower as a guide,' he told us. 'If you find yourself having to mow your lawn often, move your animals often; if the mower stays in the shed, let the animals stay in the same paddock.' Joel Salatin, the Virginia farmer who has become perhaps the most famous practitioner of regenerative agriculture thanks to a starring role in Michael Pollan's 2006 book *The Omnivore's Dilemma*, calls this 'pulsing the pastures': matching the rate you move your animals to the rate the grass is growing. Get it right and each new paddock the animals enter will contain fresh grass at the zenith of its growth. (Salatin moves his animals at 4 p.m., claiming the sugar content of grass is at its highest at this time of day.) Along with encouraging summer- and winter-active grasses, this is what my father meant, I realised, when he told me, 'You want to aim for 100 per cent green, 100 per cent of the time.' (*Of course*, I thought, when this finally clicked: *I don't want to eat something that's been in the vegetable crisper for a few weeks, so why would cattle?*)

Initially I thought Brian's mowing tip an odd metaphor – none of his students used a lawnmower much. That was why we were doing this course: to use animals as landscaping tools instead of machines. But the clever simplicity of the tip – looking to the garden for clues to how nature at large operates – impressed me, and I have thought of it often in the years since while driving through suburban Australia, where overgrazing of lawns is rampant, not by mouths but by blades, in an overzealous attempt at keeping things orderly: a domestic distillation of the desire to dominate nature that holistic management is trying to move away from.

There was a whole suite of other monitoring techniques Brian taught us, like 'adopting' a particular plant (he encouraged us to give it a name) and taking photos of it throughout the year to see how it fared under our management. He also told us to look at what comes out the back of a cow to monitor the state of what goes in the front. A sloppy frisbee of a cowpat indicated you were moving the herd too fast, as they were eating nothing but the cow equivalent of ice-cream – the lush, less fibrous green grass. A hard, clumpy cowpat that resembled a failed pavlova was a sign that they were being moved too slowly, as they had long finished off the ice-cream and were now having to settle for lower-quality dry matter – what Brian called 'broccoli'.

Using these tricks, along with a few others, I saw things on Gollion I hadn't seen before: the way a plant that has been repeatedly overgrazed will grow back horizontally instead of vertically to try to make itself harder to eat; the way recently trampled grass resembles damp tobacco, before eventually becoming incorporated into the soil beneath it. This is nutrient cycling at its most visible. I thought of these as fingerprints of management decisions, and soon I was looking for them not just inside Gollion's borders, but outside them.

Over-rested and oxidising plants, I noticed, were most prevalent on the roadside verge between Westmead Lane and Gollion's north-eastern boundary: part of Australia's 'long paddock', an expression I'd heard mentioned but not understood until now. In cities or on the edge of highways, the grass tends to be overgrazed by the local council's ride-on mowers. But on our dirt road the grass was left to grow and deteriorate, slowly turning the dusty grey of the potholed road adjacent to it. Anyone can apply for permission to graze the long paddock; all you need is a sign warning motorists that livestock are loose on the road, and to stay near them while they do. (The first time I heard this my mind pictured a bucolic, pre-industrial scene, a cowherd playing a piccolo or composing verse on cowhide with a piece of charcoal from the fire over which he'd cooked his vittles; in reality, I suspect long-paddock graziers nowadays pass the time scrolling on their phone or, if an older cohort, flapping open the pages of *The Land*.)

In the Long Paddock adjacent to Gollion I also saw African lovegrass, one of only two weeds my father thought worthy of the title, a fast-growing perennial that outcompetes other grasses and will only be eaten by cattle as a last resort. It wasn't present inside Gollion and had likely hitched a ride in a passing wheel hub. The other weed Dad didn't like, for the same reasons, was serrated tussock; he taught me how to identify both, encouraging me to run my hand up each plant from base to crown to get a *feel* for them (the serrations of serrated tussock made this memorably unpleasant), as well as remembering what they looked like until I 'got my eye in', making it sound like one of those magic-eye illustration from the 1990s.

I got my eye in for serrated tussock first, and it wasn't on Gollion or the Long Paddock that I suddenly saw it everywhere, but on the upper slope of the hill that marked Gollion's south-western boundary. The top of the hill belonged to the neighbour, and a fence cut across its upper third like a knife through a hardboiled egg.

It is rare to hear a farmer criticise their neighbour's management; keeping on good terms is a must if you want to ensure your dog isn't shot on sight if it shimmies under the fence. But 'Carey' (I still don't know his first name) was frequently cited by my dad as a lesson in how *not* to farm. His block – one large paddock stocked year-round with white Square Meater cattle – barely changed from season to season, dry year to wet year. Being set-stocked, it was grazed continuously. On one of my homework forays I skipped the fence. Bare ground was interspersed with sections where nothing but serrated tussock grew for fifty metres. Some sections were like concrete, hard and burnished. Looking back towards Gollion, you saw a canopy of different-sized grasses, in different stages of growth, young and mature.

I thought it strange that Carey hadn't ever asked Dad what he was doing differently. Even if he were embarrassed about it, wouldn't he want to copy? David Marsh, a regenerative farmer from Boorowa in the South West Slopes of New South Wales, has a theory that the ego of an Australian farmer is 100 kilometres long. They won't introduce new management techniques if it means copying their neighbour, or even their neighbour's neighbour, but they will implement something new if they saw it in another district. On a field trip to Jillamatong, a 450-hectare holistically managed cattle farm near Braidwood, the owner, Martin Royds, told us his neighbour had voiced dismay that Martin seemed to be getting more rainfall – what else could explain the disparity between his side of the fence and Martin's? We trespassed while we were there and compared the rate at which water infiltrated the soil: at Jillamatong, it disappeared within thirty seconds; on the neighbour's side, it stayed on the surface for minutes.

Maybe, I suggested, Dad could give Carey some advice. At least suggest he put a fence down the middle of his property, so that the cattle weren't grazing it twelve months of the year? My father chortled. All those denuded hills, he said. They divert the rainwater into Gollion beautifully.

I tried to school Dad with some of the techniques I'd learned at grazing school, but he schooled me with some of his own. My father had never done a course in holistic management (he didn't like the emphasis on wider decision-making; like Rudolf Steiner's more radical biodynamic teachings, he considered it 'mumbo-jumbo'), but he read widely, attended field days and observed what worked on his own farm. I learned more about how his mind worked when he read the landscape, the clever tools he used to manage it wisely. He knew about the 'dung scoring', and about the lawnmowing gauge (it had occurred to him independently). He had even adopted a plant, even if he hadn't given it a name.

Around the new millennium, he told me, he'd sown some chicory seed in the easternmost paddocks of Fernleigh. Chicory, a long-lived perennial, is about the most palatable of grasses to cattle. Every time the cattle were in those paddocks, he told me, he was pleased to see the chicory wasn't just surviving; it was spreading. As the first plant cows will go for, if he had been overgrazing those paddocks it wouldn't still be there. It was an important lesson: a paddock is not 'eaten out' when it is devoid of grass, but when its best grass – at least in the eyes of the cattle grazing it – has been eaten. The cattle will only move on to other grasses if forced to, by portable fencing, for example. Otherwise, as soon as the palatable grass begins to regrow, it will be chomped again, irrespective of how much other 'feed' is available.

I was finding 'feed budgeting' hard: looking at a paddock and telling how many days of fodder were available in it – what Dad called 'the judgement thing', the mental alchemy of timing and decision-making. It was reassuring to know that Dad used reference points in the paddock to make his decisions; he wasn't moving cattle on intuition alone, but by reading the signs in the environment.

Often when rain was forecast, Dad (used to empty promises from the nightly news) would walk over to a rock, crouch down and say to me, 'Let's see what the ants are doing.' He'd then turn over the rock and comment on how manic the ants underneath were. They always looked manic to me, but when they are *especially* manic, he assured me, it meant that rain was on the way. 'Animals know these things better than any weather forecaster.'

One day in the paddock I prodded a cowpat with a stick to see how firm it was and Dad asked me from the ute if I could see any dung beetles.

'You know, when Darwin visited Australia,' he said, 'he claimed that dung beetles had evolved since 1788.'

'To process cattle and sheep dung?'

'Yeah.'

When he got out of the ute and inspected the dung himself, Dad confirmed that these were native beetles.

'Still good,' said Dad, 'but smaller than the introduced ones.' He dug around in the dung, explaining how exotic dung beetles were introduced to much of Australia to speed up the cycling of nutrients, but he rarely saw them on Gollion. It was another example of monitoring, checking, seeing what was happening on his farm.

At grazing school, we learned how sociologists believe paradigm shifts occur. The adoption of something new – a belief, a technique, a must-have fashion accessory – is made by a tiny minority: the innovators. Next comes the early adopters, still a minority of us. Most people fall within the cohort of early and late majorities – they are open to new ideas, in time. Last to catch on are the laggards. I'd heard my mum use that expression to describe Papa: an early adopter. He planted trees when farmers were still chopping them down; he aimed for biodiversity while his neighbours favoured monocultures.

In class, Martin Royds was cited as an example of an early adopter of regenerative agriculture. By following Hunter Valley horse trainer Peter Andrews (an innovator) in slowing down the flow of water across his floodplain, he had recreated how an Australian watercourse would have functioned prior to colonisation, rehydrating his landscape and lengthening Jillamatong's growing season. He'd achieved this by building a series of 'leaky weirs', blockages of bulldozed earth, rocks and logs at narrow spots in a creek that slowed but did not stop the flow. Each was higher than the next, so that as the stream flowed slowly downstream, it did so via a series of steps. Below one of them, Martin had put the spinner from a front-loading washing machine, which the water flowed into and through. This was his yabby trap, his own innovation.

Martin's creek restoration, known as 'natural sequence farming', is rightly celebrated. It is a way of storing water in the Australian landscape that allows for less evaporation than dams and hydrates soils and pastures adjacent to the watercourse, as well as the watercourse itself. But it is no longer even the most famous chain of ponds in the region.

In 2016, writer and fine-wool producer Charles Massy was travelling around Australia, interviewing regenerative farmers. Massy, who had previously written a celebrated book on the Australian wool industry, was convinced that an 'underground revolution' was underway and wanted to connect, celebrate and document the farmers who were working with the environment. He wrote about his findings in a new book, a call to arms to address what he saw as an 'environmental genocide' being perpetrated by conventional agriculture. My father was then president of the local Landcare group and invited Massy to address a meeting. We didn't know it at the time, but Massy was taking notes.

I had been asked to address a Landcare group at Sutton, near Canberra, and I arrived mid-afternoon at the farm of my hosts David and Jane Vincent. David then took me for a drive to view some of his creek reclamation work. He runs a mob of 150 or so Angus cows, holistically grazed, and on the drive out I saw dense tree-breaks of tree lucerne, neatly trimmed back to the fence and browse-line by the last mob of cattle in the paddock.

After a drive up and down steep hills, we came to a valley, passing on the way the upstream paddock of a neighbour that drained into the valley's creek. It was a dry spring, and much of the hill country on David's block had browned off. The set-stocked neighbour's place was in far worse shape. There was little groundcover, and that which existed was a yellowy-brown. The neighbour's creek was eroded, with large areas of bare ground. David informed me that a large mob of merino wethers (castrated male sheep) were set-stocked there the year and that there were outbreaks of dryland salinity also.

Then we crested a foreground ridge so we looked down on the same creek now running through David's farm. I was met with a stunning sight. Not only were there no erosion scars or banks visible, but to a width of fully 100-200 metres either side of the creek there was a bright-green swathe of grass, including large tussocks and *Phalaris* up to half a metre high. On closer inspection, there was also a variety of broad-leaved native grasses and forbs, plus sub and white clover. The contrast to the neighbour's land only half a kilometre upstream was dramatic.

As we parked and walked down to the creek, David told me he had heard Peter Andrews talk once, and then, with a small group, he had driven up to the Hunter River and Andrews' home farm, Tarwyn Park, to see for himself. While there, he and his group also visited the farm of the famous retailer Gerry Harvey. Gerry had employed Andrews to apply his pioneering water works. David said the visit was mind-changing, because in a drought Andrews had converted the entire valley to a lush green compared with [the] neighbours' overgrazed brown. There were even green patches high up steep hillsides. I was now experiencing a similar scenario on David's place – even though he had been applying this work for only nine years.

The first thing that struck me was that, unlike on his neighbour's place above, here the creek was readily running again (the landscape was releasing stored water). Moreover, there were sizeable, weed-cloaked ponds and bunches of the large perennial grass *Phragmites australis* (or common reed), along with tall cumbungi reeds (*Typha*) and other succulent water plants. Furthermore, the creek banks had largely healed (aided earlier by cattle hoof impact before David fenced off the creek), and there was no sign of salt scalds. I then stood and looked back, outwards from the creek, and could follow how the water that was now clearly being held in the creek vicinity had begun to rehydrate the lateral landscape.

David showed me his creek crossing points (narrow trackways for vehicles and livestock), where he had built up slightly elevated banks with tyres and rocks, which impeded and backed up the water while allowing it still to gently flow through. This was a famous Andrews design called a 'leaky weir'.

> Then a sound caught my attention: the unmistakable high-pitched song of a reed warbler, high up and swaying sideways in a rocking reed bunch. I had seen many such birds in bigger reed clumps on various expeditions along river courses in the past. I was surprised, therefore, that the bird could be found in this small founding patch of reeds that David had helped to introduce. Clearly it knew there would be a much larger reed bed very soon.
>
> Later, over a cup of tea, as David and Jane checked out the bird and its call on an impressive app that Jane had on her iPhone, I pondered the likelihood that these locally rare birds probably hadn't been seen in this creek for more than 130 years. The next morning, as I left Canberra heading home, with filaments of dawn mist rising off dams and cloaking irrigated flats of lucerne beside the Murrumbidgee River, I concluded that the lovely reed warbler could only be a talisman of a watercourse and landscape function on the path to healthy regeneration.

The following year Massy's book was published: *Call of the Reed Warbler*. It became an unexpected bestseller, introducing a new audience to regenerative agriculture and propelling its shy, modest author – always dressed in a grazier's woollen sweater and collared shirt – into the role of messiah. Massy now regularly addresses community halls in the bush and writers' festivals in the city. The book's latest edition has a sticker on the cover: 'As seen on ABC TV's *Australian Story*.'

Massy's message, that a new way of farming can feed us *and* reverse the environmental damage wrought by industrial agriculture, was predictably criticised by those who farm conventionally. Fiona Simpson, president of the National Farmers' Federation, has accused him of 'drinking the Kool-Aid' and countered that if all Australia's farmers practised the techniques Massy espoused, yields would drop and the cost of food would increase. (Simpson didn't mention that industrially grown food contains fewer nutrients than regenerative and organic produce, making weight a misleading measure of a harvest's worth; nor did she specify that shit food, too, has a high cost – it's just that the bill is footed by the health system, the environment, and the mental health of the producers.)

Throughout his travels, Massy found that farmers who had made the transition to regenerative agriculture suddenly started making money: without all those chemical inputs, especially for croppers, profits began to grow. He was also repeatedly told something that has since been supported by research from the University of Canberra: regenerative farmers have better mental health than conventional farmers, who commit suicide at a far higher rate than the national average.

Massy found that for most of the farmers he interviewed, a major life shock had 'cracked their mind open' – accidents with chemicals, livestock disease, crippling debt compounded by drought. They bottomed out before they changed tack. Massy, too, had learned from his mistakes, eschewing chemicals and set-stocking only after a drought forced him to sell half his farm; one day during the drought of 1982 he sent a truckload of sheep to the Cooma sale yards only to learn that they were worth less than the freight to get them there.

My father's own epiphany had occurred in 2003, when he retired from economics and began farming full-time – in the midst of a drought. Dad thought he was 'ecologically enlightened' by then, but the severe conditions exposed how unsustainable his practices still were (rolling out hay bales for too many mouths in paddocks reduced to dust gave Dad the 'kick up the bum' he says he needed). He subdivided his paddocks and began holistic grazing, learned to brew compost tea, increased and diversified his tree planting and, most notably, was taught the fundamentals of natural sequence farming by the iconoclastic land manager Peter Andrews.

Charles Massy calls himself a storyteller, not an innovator: he considers his job to spread the findings of the farmers he interviewed. The book's greatest achievement may be in linking previously dispersed regenerative practitioners to demonstrate the existence of a larger movement: a community of farmers who love their land and who independently realised they could no longer make a living or care for the environment through conventional agriculture.

I haven't tired of telling people who recommend it to me that Massy's book was actually named after a denizen of Gollion, and Dad hasn't tired of raising the book each time we cross the creek.

'That a reed warbler?' he'll say of a bird that flutters out of our way. 'That a reed warbler?' he'll ask, when we hear a tweet, a screech, a whistle.

We haven't knowingly heard a reed warbler since Charles Massy visited – but until one of us learns to identify its call, we can never say for sure.

# 11

My father lowered the TV guide and raised his eyebrows. Inching forward on his chair and planting his elbows on the dining table, he waited, his mouth forming the beginnings of a wry smile, for me to look up from my dessert.

'Says here there's a show on soon called *The Farmer Wants a Wife . . .*'

I had made the mistake of staying for dinner.

One thing they don't teach you at grazing school is why the bachelors on *The Farmer Wants a Wife* are usually cattlemen. There are a few croppers, to be sure, parting their stalks of wheat or canola in hapless search of a mate, but it seems to me the wife drought bites hardest on the open range.

I have a theory, formed from personal experience, not as a reality TV contestant but as a participant in online dating. For a few years in my early thirties I fell into a pattern that will be familiar to many: download Tinder and its derivatives; go on terrible dates with people you'd never notice otherwise; delete the app(s).

The virtual meat-market. The digital slaughterhouse. This was not a gratifying way to spend the last evenings of my youth, although by then my father had been calling me middle-aged for years. Even when I strictly adhered to my housemate Charlotte's rules for online-dating profile pics (no fish-kissing, no bicep-flexing, no skydiving, no skiing or jetskiing, no holding someone else's baby as bait for the clucky), I always seemed to cock-block myself with the accompanying text.

'Unpaid rural labourer', I decided, wasn't going to win me any hearts. By itself, 'farmer' seemed harmless enough, intriguing even. It invited friendly queries: *Does he farm wind? Harvest organs?* The perfect icebreaker. But for some reason I felt the need to get specific. And that's when things got interesting. I acknowledge that the algorithm works in ways I do not understand, and I admit my findings are not scientific. And yet – I noticed a marked difference in romantic interest according to whether I said I was farming figs or cattle.

I reflected on how I might appear to prospective dates. Trawling through profiles, making split-second decisions, likely encouraged to be ruthless by friends and housemates and likely unfamiliar with either agricultural activity ... how would they see me?

I composed a mental list of word associations.

*Cattle farmer:*
Ruddy-cheeked
Jowly
Anglo
Oversized hat

Small mind
Hormones
Antibiotics
Climate destroyer
(Amazon) deforestation

*Fig farmer:*
Suntanned
Muscly
Mediterranean
Acceptably sized hat
Free spirit
Nutrients
Antioxidants
Climate cooler
Garden (of Eden)

The rearing of cattle – the *production* of beef and milk – is, in many ways, an ecological disaster. When a cow swallows a mouthful of fodder, it sends it to its rumen. That's what cattle are doing when they're down on their haunches: ruminating. (As they do so, they move their jaws from side to side, 'chewing their cud' – partially digested cellulose – which, after being softened by the juices of the rumen, is sent back to the mouth for a second maceration. Cellulose is tough stuff.)

A byproduct of methanogen digestion is a gas burped out by cattle: methane ($CH_4$). By some estimates, methane warms the atmosphere twenty-eight times more than carbon dioxide. One Oxford University study found that it is *only* eight times more warming. On average, a cow releases between 70 and 120 kilograms of methane annually. There are one billion cows on the planet.

Carbon dioxide itself is released into the atmosphere in vast quantities by the cattle industry, both when overgrazing creates bare ground that erodes the soil and unlocks carbon, and when land is cleared to make way for cattle pastures. In Brazil, the world's biggest beef exporter, cattle ranching is currently responsible for 80 per cent of land clearing in the Amazon rainforest; in August 2019, 26,000 forest fires burned out of control, a conflagration originally lit by ranchers.

Other pollution, in the form of nitrate runoff from feedlots, can enter waterways, creating algal blooms and spreading antibiotic-resistant superbugs. Cattle are notoriously land-hungry (three quarters of the corn and soy grown in the world is fed to livestock) and thirsty (the CSIRO estimates it takes 50,000 litres of water to produce one kilogram of beef, compared to 1010 litres to produce a kilogram of wheat, 2200 litres for soybeans, and 2385 litres for rice).

When a braver student than I raised her hand at grazing school and asked Brian how cattle can be a force for good when they are destroying the planet, he replied, 'It's not the cow, it's the how' – it's what you do with the animal that counts. All these problems, Brian claimed, could be drastically improved through holistic management. Putting cattle on grasslands; moving them through those grasslands in concert with the rate of grass growth to ensure that pasture and soil health aren't damaged; only watering them with the rain that falls; getting them out of factory farms and into fresh air where manure can fertilise, not pollute;

getting them off antibiotics and growth hormones; getting them to market when drought reduces available fodder. They are only land-hungry if we dedicate grain to feeding them, rather than letting them eat pasture; they are only water-thirsty if we don't hold the water that falls as rain in dams, and in hydrated, healthy grasslands.

Methane, a gas emitted naturally by all ruminants, farmed or otherwise, can be reduced by feeding cattle plants instead of grain. Their digestive systems have evolved to digest leaves; grain fattens them more quickly but requires more cropland and makes them gassier. Even after methane has been released, the carbon sequestered through holistic planned grazing, Brian told us, can still make a farm's emissions neutral, or even negative. It is possible to emit one form of carbon, methane, while sequestering another, carbon dioxide.

Globally, the cattle industry is so overwhelmingly bad that Brian's caveat is barely worthy of an asterisk. In the USA, the world's third-biggest beef exporter, more than 97 per cent of beef cows are grain-fed. As well as making cattle burp more, and irrespective of whether they are grown for them or us, inherent in annual crops – wheat, soy, corn, oats – is a need to break ground. But ploughing destroys the topsoil, the layer where plants grow. The UN's Intergovernmental Panel on Climate Change (IPCC) says that 50 per cent of the world's topsoil has been destroyed in the past 150 years, and that ploughing loses topsoil one hundred times faster than it's made. Even 'no-till' cropping, in which seeds are drilled directly into the ground, loses topsoil twenty times faster than it's made.

When topsoil is lost, carbon is released from the soil.

The IPCC estimates there is three times more carbon in the soil than in the atmosphere – and half as much since the advent of ploughing. Carbon in the atmosphere is bad, but carbon in the soil is good. Carbon is in the sugars photosynthesising plants put into the soil; carbon is in humus, the part of topsoil that keeps soil wet. There is carbon in the threads of fungi that hold the soil together; carbon is found inside the bodies of microbes – bacteria, protozoa, algae, nematodes – and provides them with a home and source of food. Unless you're a frugivore or getting your carbs from perennial crops like acorn or chestnut flower, a plant-based diet will require the loss of soil structure and, with it, carbon.

The best way to keep carbon in the soil, and to bring down more of it, is to maintain healthy groundcover. Repeated studies show that the fastest and most efficient way to return carbon to arable land is to return cropland to a biodiverse, year-round plant cover consistent with its climate – a forest if that's what was there previously, but in most cases a grassland. And nature, as the father of organic farming Alfred Howard wrote, 'never farms without animals'.

Pasture – cellulose – can't be eaten by humans. Which is where the ruminants come in: cows, sheep, caribou, wildebeest. When they are correctly 'pulsing the pastures' (moving through pastures at the pace at which they are growing), ruminants feed the soil by triggering new plant growth. By turning cellulose into meat, they then feed us.

Unfortunately, this process also produces methane. The good news is that methane has a lifespan of about a decade, whereas carbon dioxide can live in the atmosphere for thousands of years. As long as the global cattle herd doesn't rise – and it has hovered steadily around one billion for the past decade – the methane produced by the beef industry won't increase exponentially in the way that carbon dioxide is.

A 'climate-friendly' diet, then, doesn't fall into the dichotomy of plant-based = good, meat = bad. Holistically grazed beef can be less damaging than greenhouse-grown vegetables. The IPCC recommends eating meat from sustainable farms with low greenhouse-gas emissions. This means supporting regenerative, organic producers and sourcing locally, which reduces pollution from road, rail and air freight. What is sustainable in one ecosystem is not in another; ruminants that eat grass in low-rainfall ecosystems like Gollion produce food from soil that is unsuitable for cropping.

'It's not the cow,' said Brian, 'it's the how.'

The more time I spent working with cattle, the more I liked them. As prey animals they have near-360-degree vision, even when they have their heads down in pasture. When you walk into a paddock of cattle, they will neither flee nor approach. Instead, they stop what they are doing (inevitably, grazing) and affect a ho-hum, hooves-in-pockets posture of nonchalance. *You've got it all wrong,* their furtive eyes say. *I just like the way this handlebar of grass feels in my mouth.* Not even their jaws move.

Good cattle handling is about patience, deliberateness and consistency. Cattle are intelligent: at Gollion, they knew that they were on a circuit and would gather at the entrance to the gate leading to the next paddock in anticipation of being moved. They like routines and prefer to be moved at the same time of day. The internet is full of videos of holistically managed mobs of cattle moving calmly from one pasture to the next. Some farmers play classical music from speakers rigged up to motorbikes or utes as they approach, so their stock associate the relaxing music with feeding time. This differs from conventional set-stocking, where animals spend so long in the same paddock that by the time they are moved, they have either eroded the grass around the gate in eager anticipation or don't understand what the farmer is trying to do, requiring several people to muster them on foot, motorbike, car or even helicopter. They are basically wild animals. Gollion's cattle were generally calm but tended to rush between paddocks like children attempting to walk-run at a swimming pool so as not to incur the wrath of the lifeguard.

I learned a lot about cattle by sitting in the grass among them, pulling my knees to my chest and waiting. I learned that they have three speeds: walk, trot or rodeo. They don't 'trot' elegantly like a horse, but clumsily like a pantomime horse – a consequence, I suppose, of being shaped like a bathtub. I learned that they don't stop eating while urinating or defecating, and that their pee is so prodigious it falls in *sheets* instead of a single stream, liable to catch in the wind and form its own weather patterns. I learned that cows use their tongues to pull whole posies of grass from the ground, rather than nibbling it with their teeth like a horse does. And I learned that when entering a new paddock, before they go for the 'ice-cream', they often eat the last things I'd expect first: mallow, water lilies, bright yellow baubles of wattle flowers. Like humans, they have cravings and act on them, swishing their tails and flicking their ears when a craving hits the spot.

I learned that if you sit among cattle for long enough, they will stop acting like shoplifters and start acting like window shoppers. First, they'll ogle you from a distance. Then they'll slowly approach and see how you feel, pulling a sleeve, nuzzling an arm, even allowing themselves – if you extend an arm slowly and deliberately – to be patted. One day I reached around to pull out my handkerchief only to find my back pocket had been picked. I found the handkerchief in the jaws of a nearby heifer.

This was why Dad and I were constantly 'stock-proofing' water tanks and water pumps, tractors and motorbikes: anything that could provoke their curiosity. If you leave your ute in a cattle paddock while you go to do a job, when you return they'll be nosing about in it. If you make the mistake of leaving washing on the clothesline when the cattle are in the home paddock, they will yank it off – not to eat it, as goats would, but for the hell of it. They'll pull the cover off a motorbike seat to get to the foam underneath, and they'll chew through electric fence insulators despite the 8000-volt shock.

But the main thing I learned from watching them is that cattle are herd animals: like schools of fish or flocks of starlings, they are constituent parts of a greater whole. They have individual personalities, but mostly they move as a group. They move between paddocks as a mob, single file. In late summer, when the grass is tall, they create mazes and tracks. And they need company. Once, farm-sitting while my parents were overseas, I followed Dad's handwritten instruction to 'take the bull out of the main mob' but I made the mistake of putting him into a paddock by himself. He lasted one day before he jumped two fences to reunite with his harem. I had more success separating him from the main mob with a cow and a calf for company.

For Dad, the most important thing to understand about cattle was that 'the thing that makes you money is getting grass down their throats'. Late each summer we would get the herd into the yards, count how many animals we had ('to keep the tax man happy'), and make decisions about which to keep and which to sell. Old bulls, dry cows and 'difficult personalities' were sold at auction at weekly 'fat sales'; sometimes called 'premium' sales for marketing purposes, fat sales are where butchers buy their meat.

By contrast, a 'store' sale is a seasonal affair. This is where graziers looking to bolster their herds purchase steer and heifer calves. Most of Gollion's income came from store sales. All of the steer calves born the previous winter, and most of the heifer calves (some were kept as breeders), were weaned in late autumn 'straight onto the truck'. Prices averaged around $750 per calf (steers usually command around 20 per cent more than heifers because they grow faster, but this is reversed in the aftermath of a drought, when demand for breeders soars).

This was how Dad had built his herd over the years: keeping fast-growing, healthy and placid heifers for breeders and selling the steers. 'You want to buy cattle when everyone else is selling, and sell when everyone is looking to restock.' I filed this piece of advice in the same mental pigeonhole as when Dad told me, 'You want to aim for 100 per cent green, 100 per cent of the time' – a nice idea, but an unrealistic one, requiring exceptional pasture management. In practice, most farmers retain similar herd sizes regardless of the season and freight in hay where required.

The thing I didn't like about waving off a truckload of calves each autumn was the loss of control we had over their fate. We knew how they had been treated at Gollion, and how they had treated Gollion's pasture and soils, but what happened next? Dad couldn't say for sure where the meat would end up ('high-end restaurants, I'd say – here in Australia, but also in Japan, Korea, America'), but I was more concerned about the prospect of them overgrazing, being fed antibiotics and 'finished' on grain – a technique that fattens cattle faster but produces less nutritious meat, uses more land and increases their methane output. In Australia, grass-fed beef can be finished on grain in feedlots lots for up to sixty days without the farmer having to disclose it to the consumer.

'The whole methane thing's overblown,' said Dad when I raised it with him. We were walking among the cattle, along the track normally used by vehicles to ascend what had once been the back paddock. He was wearing one of the little flowerpot bucket hats he liked (he complained, when he wore a wide-brim hat, that he was continually hitting his head on branches and doorways). The only other farmers I've seen wearing this style of headgear are octogenarian women pulling out sweet potatoes on the island of Shikoku.

I asked him what he meant. 'There've always been ruminants, but there hasn't always been man-made global warming. Think about all the bison on the Great Plains of America before Europeans arrived – some of them weighed *two tonnes*, can you believe? The antelopes in Africa; the moose in Canada. Climate change is a result of the industrial revolution, not of burping cattle. Rumination is a natural process, after all.'

Since that conversation, I have done much reading on this topic. And I have concluded that all the contentious claims about cattle and the environment – about how much land they take up, how much water they drink, how much methane they produce – are not really problems caused by cattle, but by industrialisation and its application to every facet of life since the late eighteenth century. Dad is right: over the last 150 years, methane concentrations in the atmosphere have more than doubled, yet overall ruminant numbers have declined. Over the same period, the IPCC says, industrial activities have increased the concentration of atmospheric carbon dioxide from 280 parts per million to 416 parts per million.

The amount of methane in the atmosphere is rising – not because of cattle and sheep, but because of the methane released as a byproduct of digging for coal and extracting gas and oil. Methane concentrations are now more than 2.5 times what they were before the Industrial Revolution, although there were many more wild ruminants then. Ridding the world of livestock wouldn't stop methane accumulating in the atmosphere. If we were to get rid of all livestock, it would take about ten years for the methane they produced to disappear. Meanwhile, the unchecked extraction of fossil fuels would keep filling the atmosphere with trapped methane, and we wouldn't have ruminants replenishing the soil.

So why are 'cattle' lumped together and singled out for

condemnation? Why blame the cow and not the how? Be it pea, canola (the two main ingredients of most imitation meats), soy, corn or wheat, these are vast, annual monocultures, requiring ploughing and reliant on fossil fuels for machinery, fertiliser, herbicide, pesticide, intensive processing and global distribution. Cows eating grass grown by the sun and turning that grass into food for humans and food for the soil – it was as neat an example of my father's idea of a free lunch I could think of.

The flood of conflicting information, and the disconnection of most modern eaters from their dinner, can't help. But another reason, I think, gets to my theory as to why my eligibility on dating apps seemed to drop when I admitted to farming cows. The more I read about cattle, the more I realised they have long been a mascot of modernity and its associated environmental destruction. From the origins of private property (the 'Enclosure' acts of pre-industrial Britain) to the invention of the assembly line (Henry Ford was inspired by the slaughterhouses of Chicago); from the destruction of colonisation (cowboys, gauchos, graziers and Boers) to the terminology of the share market ('stocks', 'bull market'), cattle were always there. The 'improvement' of cattle breeds to maximise milk and meat coincided with eighteenth-century ideas about eugenics and was a foundation of early capitalism.

The word 'cattle' itself is telling. It emerged in its modern form sometime after the Norman conquest of England in 1066 from an Old French word, *chattel*, and originally also meant moveable property – including slaves. Bondage, private property, wealth accumulation: modern cattle rearing is synonymous with capitalism. And capitalism, more than any activity in the history of humanity, has created the ecological nightmare we now face.

# 12

From the air, Gollion is shaped like a plumber's wrench. Before the advent of satellite mapping, farmers paid aerial photographers to do a flyover, take some snaps, then crop to given coordinates once back on the ground. Never before had landowners seen the extent of saltpans, the way capillaries of green branched out from waterholes. Three-dimensional landscapes became one-dimensional cookie cutters, outlines as recognisable as the boot of Italy or the pubic triangle of Tasmania.

Gollion's aerial photograph was taken in the late 1980s, mounted on plywood and covered with a sheet of transparent plastic, onto which the contemporary paddock names and boundaries have been added with a sharpie and amended with each new fence-line. Where the wrench would be held, Dad has written 'Rear End'; the tool's bottom jaw is overlaid with the word 'Hy-Fer', (an in-joke: how a German friend of Dad's pronounced heifer) and on the top jaw, which juts out and forms a sharp point (there's a dogleg in the boundary fence), it says 'Swamp'.

Google has made aerial mapping services redundant. When Dad helped me erect the net above my fig orchard, we could see our progress from space one month after our progress on Earth. My father was more interested in checking which of his neighbours still hadn't put away the hay they'd baled the previous summer.

On bigger Australian farms, drones are increasingly used to check watering points, fodder availability and livestock. Gollion is too small to necessitate drones, but big enough that weather balloons occasionally crash-land, ensnaring themselves in barbed-wire fences. A balloon – this one with passengers – made an emergency landing in Ram Paddock (the nape of the wrench's neck) one autumn morning when I was fifteen, and on a spring afternoon three years later I was walking back from the chook shed when I saw the president of the United States passing overhead. It was 2003, and after touching down in Canberra that morning, George W. Bush had addressed parliament on how great the invasion of Iraq was going, laid a wreath at the Australian War Memorial that suggested otherwise, attended a barbecue lunch to which Steve Irwin came dressed in croc-wrangling fatigues, then boarded Air Force One for home. He entered Gollion airspace soon after, Harry Dog barking at the president the way he barked at possums.

Aircraft belonging to electrical companies sometimes flew over Gollion, checking the powerlines and causing the cattle to stampede. Dad was convinced they were actually government inspectors and would result in him being fined for digging too many dams per hectare.

Then there were the passenger jets: peeling west over Scribbly Gum if heading to Perth and Adelaide; north over the flatlands of Hay Upper and Hay Lower if making the short hop to Sydney. Dead Horse, the unfortunate gully where Mr Whyte disposed of his pets before leaving Fernleigh, was right under that flightpath. Sometimes, if flying over at night, crew and passengers would see a curious sight as the plane descends: a tennis court, surrounded by darkened paddocks, floodlit to a wattage normally reserved for UFO abductions.

My father's greatest feat of amateur engineering, he erected and electrified the light towers. Like Stonehenge and those blockheads on Easter Island, how he got them upright remains a mystery.

The court's design was just as idiosyncratic. Wimbledon is played on grass and the French Open on clay, but the surface of Gollion Tennis Club was best described as ant's nest. This lent a distinct home-court advantage to anyone playing. Visiting challengers would swing back their racquet for a baseline return only for the ball to hit what looked like a Viking burial mound and squirt off at a right angle. The trick was to volley whenever possible.

A bulldozer had levelled the requisite dimensions between the chook shed and the driveway, and Dad did the rest. A chain-link fence enclosed the perimeter, and two large corrugated-iron galahs perched on the fence above the eastern sideline. The latest of several original sculptures Dad had stolen the intellectual property of, he had recreated them from farm junk with nothing but a photograph, a welding rod and an angle grinder.

To sweep the tennis court after play, my father had lashed together four doormmats and attached them to a wire harness. It looked like one of those human-pulled ploughs you see in medieval paintings of peasants who couldn't afford a bullock. The loser swept the court, and Dad never lost.

My parents are competitive people. Competitive with each other (Mum's veggie output vs Dad's fruit haul), competitive with their children ('when I was *your* age ...'), and competitive with themselves (Dad could tell you how many push-ups he can do; Mum keeps a list of how many books she's read each year). Once they had their own farm, it made sense to install a tennis court. Even the Whytes of Fernleigh had one, although now it was just a levelled corner of pasture in the paddock we called Ablett.

By 2018, the year my father would turn seventy, Gollion's tennis court was barely used. It looked like a former petrol station undergoing its mandatory pre-redevelopment purgatory while the ground toxins drained away. The ants had been crowded out by weeds; the net hung flaccid, like a condom left on an oval. The umpire's chair, actually a lifeguard's chair long ago bought from a Canberra swimming pool, petulantly faced away from the court, as though *it* were the tennis brat throwing a tantrum.

My parents didn't play tennis anymore. Instead they attended a boot-camp at the local school, where, by Dad's account, Mum had become the fittest she'd been in years, and he could do more push-ups than any of the pencil-pushers thirty years his junior.

My father has always been a physical person, aware of his body and its limitations. But if he felt his power waning as he neared his seventieth birthday, he didn't show it. With the exception of the seemingly monthly skin cancer removals ('I've had so many I know what to look for now – my doctor says I'm getting good at diagnosing them myself!'), he'd managed to keep injury-free in the years since he broke his thumb. It wasn't that he didn't put himself in harm's way anymore; more that with me helping, opportunities for recklessness didn't present themselves so frequently.

There was a time I was embarrassed when people said we looked alike. I wore a cap in my early twenties to hide the precocious baldness I inherited from him; my nose, like his, is curved like a Trivial Pursuit wedge. (Increasingly, the pink one.) I think maybe it had to do with his relatively old age compared to the parents of my peers. But now even those 'young' fathers were saggy and weak; Dad still had his barrel chest and his callused hands.

I look more like him every day. In my thirties I have developed his broad shoulders and slouchy posture, hunched forward as if let slack by a puppeteer. I don't mind when people note the similarities between us now. I quite like it. I have grown interested in what interested him, as well as growing to resemble him. My sister Lucy once said I was *becoming* him. But that didn't mean I wanted to replace him.

Seventy is the age at which Australian federal court judges must retire; by then, most people with the means already have. You're lucky to be alive at ninety, lucky to be in your own home at eighty, but seventy struck me as a precipice: old enough to have grandchildren, young enough to know their names.

He *looked* seventy (the photo on the invitation to his birthday party was undeniably that of a senior citizen, hunched and surprised that his wife's 'smarty-pants' phone could take photos), but he still had what Lucy's husband Damon called 'farm strength', the ability, thanks to daily manual labour, to make hard work look easy.

Long before I'd started working with him at Gollion, he would ask me to give him 'a lift' when I'd visit the farm – in ad breaks in the football, or after a family dinner. '*Give us a lift*, will ya, Sam?' That's always how he put it. He didn't mean *to the mall* – he meant *of this trailer/timber/tractor attachment*. Better still if I had friends over: 'Will you blokes *give us a lift*?' These were opportunities for free muscle, but he always beamed when I or one of my mates complimented his fitness. If succession is a changing of the guard, a relinquishing of top-dog status, how does someone like this – someone who erected a chin-up bar in his toolshed to supplement the boot-camp class he joined in his late sixties – begin to let go? With me there to make sure he didn't hurt himself, I saw no point in rushing him into retirement until he was ready.

For Dad's birthday, my family was due to congregate at Gollion for the party. It was rare to have us in one place, and I proposed a family meeting: we'd never spoken as a family about the farm's future and what my sisters thought of my apprenticeship. In the months before the party, Mum had been looking through old albums for a photo to use for the invitation when she found a yellowing piece of paper. On it, in faded pencil and proto-cursive, twelve-year-old me had written:

> I, David Vincent, promise to pay Sam $500 if he beats you [at tennis] before he turns 13; $400, 14; $300, 15; $200, 16, in one set played to advantage.
>
> *David's signiture* [sic; here Dad had signed]
>
> November 24, 1996

I have no memory of writing the contract and I never challenged for the money. I suspect it was Dad's idea: a gentle nudge of his mummy's boy towards manhood. But I've never been competitive like that; at twelve I knew I wouldn't beat him at sixteen. The contract didn't specify, but I wondered what the terms of recompense would be in 2018. Would I have to pay *him* if he won? Twenty-one years had passed since the challenge. Neither of us had picked up a racket for years. But he was undeniably still the top seed.

Usually the eldest son takes over. If the holdings he stands to inherit are vast and profitable (often signified in Australian farm names by the presence of the word 'Downs', 'Springs', 'Creek' or 'River'), he will know this from a young age, coasting through whichever boarding school to which he is dispatched, then relying on the connections he makes there to work as a stockbroker, lawyer or banker before returning home to claim his birthright.

According to one study, in only 10 per cent of Australian farm successions does a daughter take over. The archetypal characteristics of the Australian farmer – physically strong, independent, domineering – continue to be associated with masculinity; in conservative rural Australia, farming children are socialised into traditional gender roles. This is despite half of agricultural degree graduates being women and half of Australia's farming income (often in the form of off-farm income, an increasing necessity) being earned by them. Before 1994, Australian women couldn't list their occupation as farmer in the census. They had the options of 'silent partner', 'domestic', 'help mate' or 'farmer's wife'.

I was the only son, and I was taking over. But this development was a surprise to my sisters, not least because Gollion was never regarded as something worth passing on. Neither had I shown an interest in farming when we were growing up, nor an aptitude.

We are a toilet-seat-put-back-down family. Four against two, it was always going to be so. Back when we all lived at Gollion, my father more a benign mystery to me than a revered role model, he singled me out for a series of man-making ventures he too had once enjoyed. I lost interest in Cubs after two weeks and refused to return; the toy band-saw he gave me one Christmas was set up in the shed next to his big boy's version but remained unused.

There was a 50cc Yamaha PeeWee he borrowed from a friend for me when I was ten, one of those itty-bitty motor-bikes ridden by bears in the sort of circuses now banned in most countries. Today I regard motorbikes as tools like any other and know that my Papa, universally respected as a farmer, once took his to the mechanic because he didn't know you had to turn the ignition with a key. But back then I was afraid – of hurting myself, but also, I think, of playing a role I felt unfit for. We had to pretend the engine was running so he could take a photo of me for his own father. I grabbed the handlebars, crouched and looked ahead to a non-existent open road. I wonder if my grandfather noticed that the kickstand was down.

My father and I both loved Australian Rules football, so there was that. But even my choice of team wasn't as manly as his. Essendon, the team he had supported since boyhood, won the grand final the first year I took notice; my team, Geelong, was once nicknamed the 'handbaggers', and throughout my childhood the TV commentators called them 'bridesmaids' for their habit of perennially losing grand finals.

I joined a football team at his urging, but I was slow, cowardly, and only acquired ball skills after I stopped playing. Dad's boundary-line coaching consisted of shouting 'HANGONTOEM' when I'd drop a mark; he said his father used to shout the same thing at him. Mum told me that soon after they were married, Dad told her that the day he proudly announced to his father that he had made his high-school swim team, the old man replied, 'Well, don't expect to make it again next year.'

Watching football on TV was what we did together before we had farming. Dad would point out which player was from which agricultural region; who was the son of dairy farmers; who drove combine harvesters in the off-season. At half or three-quarter time he would retrieve his flowerpot hat from the hook outside the side door and announce he was going off to fix a pump or finish a fence and would be back soon. I'd sink further into the couch and continue watching the broadcast, even when it was just ads.

There was no football to watch in summertime, and until I picked up cricket as a teenager the closest thing we had to father–son bonding in the summer was me looking on in astonishment as he casually wiggled out of his jocks on the swimming pool bricks, revealing a sturdy pair of white bum cheeks and sometimes more before he'd change into his 'togs' and crash into the water. *So that's what a man looks like.*

Mum has told me Dad was overcome with excitement when he learned his fourth child was a boy. But once I was too big to bounce on his knee while he drove his tractor, I can't remember him ever asking if I'd like to help on the farm. I'm sure he did and gave up asking after the first time I said no. I was more of a nature boy than a farm boy. I can see now the bushland 'bases' I built back then were the foundations of my love of Gollion today. They created an attachment to place, a connection that deepened when I learned to care for that place as my father's farmhand.

My mother was more successful in getting me to help outside. We'd plant vegetables together and rake up leaves for the compost heap. 'Pottering', she called it. Mum condoned contact sport but discouraged violent cartoons and banned toy guns. G.I. Joes and green plastic soldiers were verboten, although for some reason I was allowed to take cowboys and Indians with me on my forays into the bush. She always liked it when I let the Indians win. As with those screenings of *Dances with Wolves*, cowboys in those days seemed like relics of the Old West, not us, here, on our own frontier. She seemed to have no awareness that those Indians would've seen us as cowboys too, and would have tried to split our heads open with their little plastic tomahawks given half a chance.

On the day of the party, overcast and unseasonably cool, we reverted, as adult siblings who don't see each other very often do, to our childhood roles. Eve, the eldest, socialised with the family friends who were earliest to arrive; Mum was in the kitchen, frazzled at the tasks ahead; Lucy, a psychologist, calmed her down. As the youngest, I was bossed around by my sisters: I hadn't made enough pizza dough for the expected sixteen guests; the baked vegetables didn't go into the oven early enough. Only my sister Sarah couldn't fly in from Perth, having just returned from overseas and about to begin the practical element of her medical degree. I slept off a hangover in the hours before the party under an apple tree in the top orchard, inching up the hill several times over an hour to keep in its shade.

We sang happy birthday in the living room, surrounded by antique clocks that hadn't kept the time in years. I led the chorus of 'Hip, hip, hooray!' The candles – not seventy, more like twenty – were deeply set in a triple sponge cake, made by Lucy in honour of Dad's mother, whom we called Nana and who, our father always said, made a 'mean' sponge. This one had passionfruit icing, yellow and dotted with black pips. Lucy's daughter Quincy and Eve's son Billy sat at the table and eyed the cake the way the hyena eyes a child. There was a pavlova on the table too, with peaches from the top orchard ('I can't eat stone fruit I haven't grown,' Dad said, when asked if he had a dessert request).

When it came time for Dad's speech, first he thanked everyone for coming, especially those who'd travelled from afar. Then he paused and looked around; it was clear he hadn't thought about what he'd say next and was winging it.

'Well, I've had a pretty good run. Stable upbringing. Loving parents. Good education. A beautiful wife who's been by my side for forty-six years now – and she seems to be getting more tolerant with age. What have I forgotten?'

Billy, aged seven: 'That Billy's the best!'

Quincy, aged three: 'No, I'm the best!'

Laughter. Dad smiled. In the 'local village', he continued, he and Mum go to boot camp. 'And at our little class we're meant to do push-ups in little numbers – ten at a time, kinda thing. Well, I think that's a bit stupid, so I tend to stay down on the ground and do as many as I can while I'm there.' The teacher, he said, was mightily impressed by this dedication and often singled him out to new members of the class, saying, 'One of the fittest members of my group is also its oldest.' To this, Dad said, my mother 'never failed' to reply, 'He's *sixty-nine*, you know?' Dad said he wondered whether Mum liked saying it to get him to slow down. At this point I caught Lucy's eye. It said, *He's rambling.* But then he reached his point.

'So anyway, we do our boot camp class in the local school, and the classroom is covered with Post-it notes, messages about how to live your life – with integrity, honesty or whatever. And it's a dynamic thing. New messages are always appearing. I've recently noticed one, and it's really stuck with me. It says, *Be yourself. Everyone else is taken.*

'I'm not sure what it is about this particular message, but I think about it while I do my push-ups or whatever. I've realised I've spent a lot of time trying to tell my kids to be the way I want them to be, rather than letting them be themselves. They've rejected this advice – two of them have even written very clever, smart-alecky articles about it. [One of those articles will turn into this book.] So from now on I'm going to be myself, because everyone else is taken. What this means, I guess we'll have to wait and see. Look out.'

Sometimes on farmhand days I rode my bicycle to Gollion. I preferred the way the country unfolded when I rode to the way it blurred by in the car. Roads that had been unpaved when I was a boy were now largely sealed.

Over the hill that marks the border between the ACT and New South Wales the dumping started: grass clippings and washing machines, bottles and prams. But slowly the bush asserted itself with more trees and fewer houses, cattle grazing the roadside with their heads through the fence, galahs passing overhead. Like Jehovah's Witnesses, they always work in pairs.

Hitting the dirt, my tyres caramelised and I got my first glimpse of the farm – one lone yellow box on our biggest hill – still ten minutes away. The tree itself is on a neighbouring property (the boundary fence skirts the summit), but its reach is so wide, and the hill so steep, that the morning sun casts its shade onto Gollion. It has a distinct shape, this tree, forking early from its trunk like a head of broccoli. At night, viewed from the back gate, it is silhouetted against the artificial light of Canberra. You can see it from the summit of the city's Mount Ainslie if you know where to look.

It became routine to lock my eyes on the tree to avoid concentrating on the last climb, and as I did I invariably thought of a photo on the noticeboard in Gollion's kitchen. I am out of shot, busy being a baby. My three sisters are early primary schoolers, grinning through missing teeth, each holding a half moon of watermelon. A background detail I had never noticed until I started working at Gollion and paying attention to its landscape: the old yellow box. More than three decades had passed since the photo was taken, but the tree looked just the same.

Other photos from that time captured a landscape so transformed I now can't recognise it. Being the youngest by five years, I have no memory of the clapped-out sheep farm my parents bought. That Gollion, the original Gollion, belongs to my sisters more than me. Before I could walk, they were climbing its trees and exploring its hills, playing in craters made by bulldozers, then swimming in those craters when they filled with rain to become dams.

My eldest sister, Eve, was a bookish, dreamy kid; she had secret places in the old back paddock where she entertained imaginary friends. One of those places is now in the paddock we call Siphon, a steep, south-facing hillock with a troll's staircase of granite boulders. It is cool and damp year-round; maidenhair ferns grow among the boulders. When it snows at Gollion – once a year when I was young (including the day I was born); twice a decade now – this hillock is the last place for it to melt. Eve has carried the farm with her into adulthood and visits whenever she can. My other sisters aren't so sentimental.

Lucy and Sarah spent their country childhoods destroying things more than growing them. As well as setting fire to tussocks and the occasional hillside, they broke windows, their toys and their bones. I still get them mixed up on the phone, but when they were nineteen Lucy became a centimetre shorter than Sarah when she landed on her back during a snowboarding competition.

It would be unfair to say that my twin sisters showed no interest in farming. But as the rebels of the family, the crop they chose to grow in the bush behind the glasshouse was both illegal and – the greater crime in my parents' eyes – a monoculture.

Every Wednesday in early high school, a teenage neighbour would drive to Gollion and give me lessons on an untuned piano in the corner of the living room. In the lead-up to what would be my final lesson, my sisters offered me chocolate-chip cookies they had baked themselves. Not questioning this uncharacteristic display of charity, I greedily scoffed two. All I remember of what followed is repeating the same piece of music over and over, even though there wasn't that little sign with the lines and dots to indicate I should, and how very scary *The X-Files* was that night. They maintain they had made two batches of cookies, and somehow I accidentally ate some 'special' ones. Without telling him why, they called Dad and told him to pick up pizzas on his way home from work. 'Bring an extra one,' they said. 'Sam is really hungry.'

Children aren't left home alone much anymore, especially on farms. But even years before the cookie episode, before the twins joined Eve in high school, it was just me and them in those long hours between the end of the school day and Mum and Dad coming home from work.

I date my fear of snakes from this time. Land sharks, I call them. When I was in kindergarten, it was a ten-minute walk from the house to the end of our driveway, where the school bus would pick us up and drop us off. I don't think I'd ever seen a snake in that patch of grass, but I knew they were out there, lurking, and I'd get anxious as the walk wore on.

One day I reached the house to find my twin sisters had locked me out. I screamed and banged on the side door, but Lucy only opened it enough to hand me a bowl, telling me I wasn't allowed inside until I had filled it with strawberries so she and Sarah could make milkshakes. We *had* seen a brown snake in the strawberry patch earlier in the summer, but what choice did I have? I filled the bowl with strawberries, blubbering the whole time, flinching at every rustle in the grass. A few years ago I told this story at Lucy and Sarah's fortieth birthday party. They still thought it was funny.

The day after Dad's seventieth birthday party, we held our family meeting, under the garden's crab apple tree. It was a Sunday morning, hot by 10 a.m. Evenly spaced around the table were me, Dad, Lucy and Eve. I texted Sarah in Perth so she could call into the meeting. Between us was a platter of Aldi croissants, with the addition of Aldi ham and Aldi cheese, warmed in the oven. The ham was still a bit frozen.

Farm succession can strengthen families, but it can also tear them apart. The man who made our hay lived on a property near Sutton carved out of his parents' farm. His sister lived on the neighbouring property, carved out of another part of it. They didn't speak. My mother knew of other people who were estranged from their sibling(s) over farming wills. Either the anointed successor was thought the wrong choice by their rival siblings, or else no successor had been anointed before the patriarch died.

My father once told me that when intergenerational transfers fail, it is 'usually because of the women': brothers who grew up on a farm and take it over may well get along, but 'their wives and girlfriends tend to fight.' He could cite no evidence for this claim, although he had been watching the TV drama *Big Love* at the time, about a Mormon polygamist in Utah and his feuding wives. Not, I countered, because the globalised economy forces small operators to compete with bigger, cheaper and more 'efficient' rivals in a system where middlemen – agents, packers, freighters, retailers – take a cut, making many family farms less viable for successors than they were for previous generations? 'That too,' he conceded.

Faced with falling incomes and climatic uncertainty, many farmers now actively dissuade their children from taking over the family farm. There's always a developer or agribusiness ready to buy them out. Regenerative agriculture, with its lower barriers to entry (fewer inputs, smaller scale), is being taken up by women at a higher rate than conventional agriculture – they are buying their own farms instead of hoping to inherit them.

I had no brothers to compete with. But what did my sisters think of Gollion after all these years? Did they want to sell? I could not afford to buy them out and likely never would; Gollion is not prime agricultural land, but there are few tracts of acreage this size within such a short commute from the nation's capital. In the previous decade, several of the remaining large farms in the district had been subdivided into residential blocks. I knew that if we sold Gollion and quartered the winnings, I would make more money than I ever would farming it. But the thought of Gollion's hills one day being covered with driveways and street lights, all coming on at once at dusk, filled me with dread and sadness. I was nervous. I knew that I was the only one who wanted to farm Gollion, and it pained me to think I would likely be the last person who will. But I also knew that whatever came of the meeting, family was more important to me than farming.

I'd proposed this meeting, I began, because I wanted to start a conversation about the future of Gollion; I wanted to hear what my sisters thought about me spending more time here, working on the farm. I said that I hoped to manage it one day, to live here again. That I had grown to love the work and the farm but was struggling to set a course for the next period of my life. I said I didn't want to force Mum and Dad out before they were ready, but nor did I want to be a university research assistant indefinitely. I said – perhaps unnecessarily; I was already panicking – that I felt like twenty years ago I could've made a career as a writer, a journalist, but that those jobs had gone and I needed a fallback. And although agriculture was risky in its own way, I felt a strong pull to care for Gollion and grow food in its soil.

Lucy said, 'As far as I'm concerned, this is between Dad and Sam.'

Eve took over. 'Thank you, Sam, for calling this meeting. It's very proactive. But Lucy, this isn't between Dad and Sam. There's no way this farm would be the success it is today without Mum's work in raising a family, helping out and allowing Dad to pursue his career while running this place.' Where was Mum? Inside, cleaning up the kitchen. We called her outside to join us. When she did, she wouldn't sit down but stood at the end of the table, nervous.

When Dad spoke next, it was clear he and Mum had discussed what would be said. He told us that when he and Mum died, the farm would be left to us children equally, split four ways. His will, he told us, would set out how the farm was to be valued, 'so that none of you kids can do it in a dodgy way and rip off your siblings.' I knew that a reduction in Gollion's current size would make its viability as a farm business less certain, and that if one of my siblings chose to cash in their share, I would be unable to buy them out. That's what I thought about as he spoke. I realised I had made no eye contact with him when I had said my opening piece and was now only furtively looking his way.

One of my sisters said she could give no guarantee that she would not sell her share of the farm once Mum and Dad died. She talked of the stress of paying off a mortgage in Sydney, of how hard she worked. Dad said he hadn't known she felt under such pressure. 'You know, pet, I have $100,000 worth of cattle walking around out here that we could sell to help you out.' My sister said she had *no idea* they were so valuable (her eyes opened wide). This surprised me, and it dawned on me that perhaps my siblings didn't understand the economics of the farm as closely as I'd come to. (Another time, she had assumed I'd want to focus on 'hipster fruits' once I took over Gollion, rather than cattle – not realising that I'd need to do both to scrape a living.) She declined Dad's offer and thanked him for making it.

Mum, still standing, said that before she and Dad died they might decide to sell some of the farm off and give the proceeds to us kids, especially if I were struggling to make a living off the land. Her voice wavered, and she added that her brother, my uncle Michael, wanted her to remind us that our grandparents went into debt to buy out his siblings, a burden he found so great he eventually sold up and retired. It could be even harder for me, she said: 'Sam might find it hard to convince a woman to live way out here.' I said that I hoped that wouldn't be the case. It was hard, I thought, to make plans for the future based on the reality of the here and now.

Dad spoke again, pointing out how high the rates would be once I was living at Gollion because of its proximity to

Canberra. 'That'll be your biggest problem. But I don't have to worry about that.' He suggested I build 'hundreds' of stables and charge 'city people' to keep their horses here. Was he joking? We discussed which part of the farm we'd sell off if we had to; parts of Fernleigh were closest to Canberra and held the least sentimental attachment for me.

We charted a roadmap to succession. Within five years, Mum and Dad would be living full-time at their holiday home on the New South Wales South Coast; I would be living at Gollion, which I would manage. Mum and Dad would visit often to help me with farm work. When they did, they would stay in a self-contained part of the house, where I had lived as a teenager. When they died, assuming none of it had yet been sold off, Gollion's ownership would be split four ways. Lucy looked bored; Eve said nothing. But I knew she loved Gollion as much as I did, and that she would be sad if or when it was sold. Her two lads, Ned and Billy, loved exploring its paddocks and her partner, Shane, regularly came and shot kangaroos for meat and hides. (He stayed inside the house during the meeting, along with Lucy's husband, Damon. Spouses had been barred from the discussion, said Mum, because, as I was single, their presence would place me at a disadvantage.)

As we cleared the plates and left the crab apple tree, I felt flat without knowing why. The meeting had gone well, at least as well as I'd hoped. But it was exhausting, all of these milestones. First Dad's seventieth, and now this. I was still the youngest of the family, but I felt mature in a way I hadn't before. I look back on that weekend as an important step into adulthood. I had been entrusted with a great responsibility. Whether it would prove a burden or a liberation, I didn't know.

Afterwards, Dad and I took some of the party guests on a drive to the creek. He and I sat in the cab of his 'good' ute (a rare off-road foray for it); my uncle Michael was squeezed in beside us. Crouching in the ute's tray were Michael's partner Caroline and Eve's boys. Eve had left on foot already and would meet us part of the way there. We drove in silence up into the hills, until the part of the track in Scribbly Gum where Dad called to those outside to duck their heads to avoid getting hit by overhanging branches. Michael suddenly spoke, as if halfway through a conversation. 'That's the only time I saw Dad cry. The day I left to become a stockbroker. He knew that the farm ended with him.' Koonje, Papa's farm in the Western District, was sold while Michael was living in London. I was embarrassed that Dad was beside us to hear that.

Neither of us responded to Michael, and I was glad when we arrived at the creek and Dad could revert to the spiel he gave visitors unfamiliar with the restoration work he had done: the leaky weirs; the stands of cumbungi reeds where Charles Massy had spotted the 'little brown bird' that had given Gollion a modest degree of fame. He told Michael and Caroline to compare the dry paddocks we had driven through to the lush collar emanating outwards from the creek. He snapped off a piece of red grass, the native summer-growing perennial. 'The early settlers hated this stuff. They got that wrong.' Michael asked Dad about a few other grasses, and when we returned to the car he got into the passenger seat, too busy talking to concentrate on what he was doing. 'You can drive, Sam.'

I started the engine and waited for Michael to clamber up onto the ute tray, where he was joined by Caroline and my nephews. Dad was looking at the green swathe of creek bank, thinking. He shut his door and turned to me. 'It's aaaaalll about moisture.' And of course, he was right.

PART IV

# Settler

The land, the land, the 'man on the land'. They said it in pubs, in political harangues, in newspapers and in magazines. I'd begun to think that one lived in America, in Europe or in Afghanistan but on Australia.

**Anne Baxter, *Intermission*, 1976**

But didn't you know? Australia doesn't like people.

**Patrick White**

# 13

In the winter of 2016, I was splitting firewood one morning when I received a call from my friend Dave Johnston, an Aboriginal archaeologist and heritage manager with thirty years of experience.

'Vinnie, good news. I just spoke to Matilda House. She's keen to visit the site.'

'Who's Matilda House again? Some kind of bigwig?'

'Shit yeah. She's the one who greets the *Queen* when she comes to Canberra!'

Aunty Matilda House: I knew her by sight, if not by name. When the Ngambri elder led the first ever welcome-to-country ceremony at an opening of federal parliament (on the eve of Kevin Rudd's 2008 apology to the Stolen Generations), she did so wearing a cloak of thirty-five brush-tailed possum pelts.

It was cloak weather on the August morning a few weeks after Dave called when Aunty Matilda visited Gollion, but that day she wore a high-vis hoodie and a woollen scarf. After tea, cake and the revelation that she'd declined to open the forty-fifth federal parliament the following day 'because Pauline's there', Matilda, Mum and I got into Dave's LandCruiser.

We drove into the hills, thawed frost slippery under our tyres. Our first stop was the ancient yellow box gum tree with a scar at its base, thought by Dave and Matilda to be left over from a coolamon (carrying dish) carved perhaps 200 years ago. But we admired it from a distance. The scar stood four metres outside our farm's southern boundary on the property of a grazier who, when Dad phoned her, said she didn't want any 'Aborigines' on her land. It's a stark example of the mistrust that still exists between many farmers and Aboriginal people; Matilda told Mum ours was only the second farm she'd been invited onto.

There are more than 65,000 Aboriginal heritage sites – from middens to missions – registered in New South Wales, but in 2016 there were only eleven Aboriginal Places on private property. These receive legal recognition and protection from the state minister for the environment, as they are declared to have or to have had 'special significance with respect to Aboriginal culture'. An Aboriginal Place is one that 'can have spiritual, historical, social, educational or other significance or could have been used for its natural resources'.

We were taking Matilda to what we hoped would be the twelfth.

Four months earlier, I had joined Dave as he undertook an archaeological survey of Gollion. In his job with the New South Wales Office of Environment and Heritage, Dave was tasked with finding and documenting new sites of Aboriginal cultural significance. We knew each other socially (he was dating a friend of mine); he knew my family owned a farm (his daughter lived down the hill at

Westmead Park). When he asked me at the pub one day if he could come out and see if he could find anything, I asked to tag along. If I understood how Aboriginal people had once lived in this landscape, I figured, my capacity to care for Gollion would be enriched. That was the way I thought of it: *their* past informing *my* future.

It had been a dry February, March and now April. The foodbank of summer grass, having turned yellow in November and stopped growing, was nearly eaten out; the 'autumn break', the first significant rainfall event of that season, which prompts a flush of cool-season plants before the onset of winter proper slows growth to a near halt, had not yet occurred.

With no cool-season grasses yet growing and the warm-season grasses mostly grazed, Gollion resembled a bowling green in a town where water restrictions were strictly enforced. This, Dave said, is why archaeologists love droughts.

At least that made one of us. I was embarrassed to show Dave around with the farm looking so dry, but he couldn't contain his excitement. 'Buddy, this is the ab-so-lute *PERFECT* time of year for this,' he gushed as his LandCruiser crested one naked hill after the next. Uncontained excitement, I would learn, was Dave's usual state of being.

I suggested we first drive up the hills of the erstwhile back paddock, and, once over the ridgeline, park in Scribbly Gum paddock and walk from there to the creek. This was the skinniest part of Gollion – the handle of the plumber's wrench – and that way we could methodically walk through each paddock, starting from the back of the farm and working our way to the front.

It was warm as well as dry; still and clear. We could see far across the landscape in addition to everything on it. As we rubbed sun cream onto our arms and started walking away from the car, I realised I didn't know much about Dave. Who were his mob? Where was his country?

'If they made a movie of my life they'd call it *No Land's Man*,' he replied cheerfully. 'I don't know who my mob are, you see!'

It didn't exactly sound like the feel-good-hit-of-the-year to me, but Dave was upbeat as he shared the plot. Born in Brisbane in 1968, he was put up for adoption. ('My [biological] mum is white. She got pregnant to a blackfella and thank *God* her family were good Catholics, they didn't decide to bump me on the head. So the fact I'm here, buddy, is a miracle!')

Dave's adoptive father was a lighthouse keeper. The family moved up and down the Queensland coast, including stints living on islands where they were the only humans. It was this *Round the Twist* childhood – left alone to explore caves and cliffs, a sandwich and a thermos in his backpack, expected home only in time for dinner – that set Dave on his course. He was the first Indigenous Australian to graduate with a degree in archaeology with honours, and has now worked on over 2000 heritage projects, from Point Nepean to Cape York.

We walked west through Scribbly Gum, each of us on our own track made by the tyres of Gollion's farm vehicles. Dave demonstrated to me 'emu bobbing', the process whereby archaeologists walk three metres apart (the limits of peripheral vision) and hunch over the ground as they slowly move across it.

I had recently read that one *billion* people may have lived in Australia cumulatively, and now I believed it. Dave was picking up 'waste flakes' what seemed like every few minutes, pink and white fragments of quartz that are left over when stone tools are made.

Dave is handsome, with an open, friendly face that invites conversation. He loves to talk and has the showman's tic of addressing a singular audience by the plural 'folks'. He explained to me ('so, folks') the process of lithic reduction, where a stone 'core' is hammered with another stone to create shards, which are then shaped into a desired blade, point or axe. Dave made a fist and told me to imagine the valleys between each knuckle were the empty space in a tool that are left when a waste flake is chipped away. We hadn't even left the track yet.

The scar tree on the neighbour's place, so named for the mark left in the trunk where wood or bark has been removed by a stone tool for a canoe, shield or carrier dish, was his first significant find. Descending a gentle slope, his view unencumbered for hundreds of metres, Dave saw it immediately: a dark vee of old wood, surrounded by a trunk of cream, pink and peach. But curiously, the scar was more evidently a scar – a recess, something missing from the original – from a distance than close up. It was, Dave told me, perhaps the oldest scar tree he had ever seen. ('This is some serious shit, buddy.')

The fence-line there is unusual, a dogleg that circumvents the scar tree, as if the boundary was made with the tree in mind. Dave thought the tree likely *was* the boundary, explaining that early settlers favoured landmarks to demarcate private property before permanent fencing was constructed. (This one seemed a curious choice: using an obvious sign of Aboriginal occupation to lay your own land claim.) The boundary fence had recently been rebuilt by a contractor and was now strained so tightly that when we climbed back through it, having taken a look at the scar, its wires sung.

We spent five days over a few weeks of that dry autumn together at Gollion, heads down, 'walking the country'. Dave found scars made by stone tools and scars made by metal tools where frontiersmen had nicked trees to indicate property ownership or adzed timber for building materials.

Each time he found something interesting, Dave sat cross-legged and entered the details in a notebook, logging the GPS coordinates with his phone. Every Eureka moment was classified with uncensored enthusiasm.

'Buddy, this is a fucken' volcanic tuff!'

'Feel that, folks – a hematite rock!'

'Fuck me, a knapping bay!'

In said bay (a kind of stone-tool workstation, identifiable by its high concentration of waste flakes), we got down on our haunches to look for tools. I nonchalantly picked up a rock shaped like those menhirs Obelix was always lugging around on his back, albeit small enough to carry with one hand. I showed it to Dave.

'This anything?'

He smiled and tilted his head back. 'That's your fucken' hammerstone, buddy!'

Some days we brought a picnic lunch with us; others we returned to the house to eat. One day we pulled freshwater mussels out of the creek, prised the orange meat out of their shells and cooked them in the embers of a fire. ('I think, buddy, in restaurants they keep them in a tank of clean water so they don't taste so gritty.')

Dave taught me to see the country anew, to appreciate the continuity of human experience through time and space. A nice view in 2016 was a nice view in 1816. The base of a large tree was a good place to find artefacts, because this was where passers-by – white swagmen, Ngunnawal hunters, that recent fencing contractor – would've sought shade while they rested from the midday sun. Sometimes, said Dave, artefacts showed up in seemingly random places, as if 'somebody stopped to take a leak but then forgot to pick up his gear again.'

When I worked at Gollion I began to imagine Aboriginal people superimposed onto the landscape, especially, for some reason, when I was riding a motorbike through it. Sometimes they were carrying their quarry back from a hunt. Sometimes they were sitting by hilltop fires. And whenever they threw on a new log, I saw the sparks enter the sky.

In one month with Dave I had learned more about the Aboriginal history of this district than I had in the previous thirty years. I finished high school in 2002, two years after a crowd of 250,000 people walked over the Sydney Harbour Bridge in a public call for meaningful reconciliation between Australia's Indigenous and non-indigenous peoples. But on the Monday morning after the march we learned nothing of what needed reconciling.

My primary school was even worse. By the mid-1990s the Aboriginal and Torres Strait Islander flags were flying above the playground (below the Australian one, of course), but that was about it. I have no memory of Indigenous students or Indigenous events; we didn't even do 'Aboriginal' dot painting with cotton buds on bark, as I've since learned was a staple of Australia's education system elsewhere.

We knew the names of our region's first white settlers – our very own Jebediah Springfields – because they were the school's athletics houses. But only in writing this book have I come to learn that the road we travelled to school on, Tallagandra Lane, is named for a Ngunnawal word meaning 'many crows'.

Why didn't we listen? Why weren't we able to hear?

When I later questioned Dad, who seemed strangely uninterested in Dave's survey – and, before Dave reassured him that an Aboriginal Place listing would not result in him losing the title deed, outright hostile – and had opted not to accompany us, if he ever thought about the Aboriginal dispossession of what is today Gollion, he told me that he'd never thought about Aboriginal dispossession *at all* 'until me and Mum went to see that play, *The Secret River*.'

*The Secret River*, a 2005 historical novel by Kate Grenville about a freed convict who realises his land claim on the Hawkesbury River isn't his to settle, was adapted for the stage in 2013 – just one year before I started working with my father at Gollion.

'Never? Not at school or uni, or even during the debate around *Mabo* in 1992?'

'Never.'

My sister Eve speaks about the uneasy sense of absence she felt as a teenager in the old back paddock, her curious mind opening to Australia's foundational sin, the violent theft of the land that has enabled my family and our society to enrich itself. But my father saw opportunity in this landscape, rather than absence.

Maybe it was because my family had moved into the district relatively recently. The process of dispossession had been completed before we arrived; the last Ngunnawal, Ngunawal and Ngambri families that hadn't died of smallpox had been herded into missions, fenced out of grasslands along with the kangaroos. Did my father think there were no skeletons in our closet?

As recently as the 1940s there were Aboriginal people living on Papa's farm Aratula, on the Murray River. In a letter to Eve in 2002, he wrote:

> There were two large mounds (about as big as a house) alongside two picturesque lagoons, backwaters of the creek. These were Aboriginal campfires where fires for cooking and warmth were kept going day and night.

And in Gippsland, in what is now my mother's cousin's farm, the 1843 Warrigal Creek massacre – Victoria's bloodiest, when white settlers murdered some 150 Gunaikurnai people – took place twenty metres from what is now her front door, in a house built twenty-two years afterwards. It is within touching distance, this destruction.

One day in the second week of our survey, I took Dave up the ironstone outcrop that my family called Bald Hill. Something caught his eye.

'Fuck me, Vinnie, a thumbnail scraper!'

Blunted with age, and the size and shape its name suggested, the stone tool had been fashioned long ago to mine ochre the colour of turmeric from the hill it now lay upon; further investigations by Dave and archaeologists from the Australian National University found hunks of loose ironstone from which the ochre had been mined, gouges running helter-skelter across each rock's surface like knife trails in butter.

Ochre was traditionally rubbed onto the skin during birthing, coming-of-age and bogong moth–feasting ceremonies. It was also used to welcome visitors onto country – country that has been inhabited for at least 21,000 years. Most of the region's ochre quarries are now buried under Canberra's suburbs, so, after finding this one, Dave encouraged us to seek its listing as an Aboriginal Place.

We wanted to do so in collaboration with the quarry's traditional owners. The first visitor was Carl Brown, a softly spoken Ngunnawal elder whose gleaming white sneakers were soon caked with red mud. In a hands-on-brims westerly wind, Carl carefully listened to Dave explain the archaeology of the site and ran his fingers over a piece of rock with veins of ochre. After examining the gouges, which Dave hypothesised were made by the thumbnail scraper (Carl thought they looked more like rotten teeth or corrugated iron), he became reflective.

'I think about all my family before me and how things were, and I get really upset about how people were treated, you know what I mean? I have a really strong belief in my culture. [This is] such a nice place. I'm proud to be here today.'

When Carl was a boy, he told me, he used to go rabbiting without permission on a farm near Yass. One day he was caught by the farmer, who chased him off the property on horseback. On another occasion, this time an adult, he was on a farm near Canberra's Black Mountain. Like today, he had been invited by an archaeologist. But the farmer still asked Carl what he was doing there.

'He [the farmer] said, "Aborigines are coming out of the woodwork nowadays."

'I said, "I've always been Aboriginal." I was so angry, I wanted to punch him in the face. He was a racist cunt.'

Carl told me that if farmers bothered to engage with his community, they'd realise 'we just want to have a yarn'. Out of the wind, in the dappled shade of some nearby scribbly gums, Carl pulled out his phone and showed me a black-and-white photo of his great-great-grandfather: a muscular, bare-chested man in his thirties with a full beard and twirled moustache. A half-circle brass breastplate hung from his neck; the caption, painted in white cursive, read 'King Andy Lane'.

'Andy's a whitefella name,' said Carl. 'Our people called him Audi.' As Carl explained, the term was both a name and an adjective. 'If you were *audi,* it meant you were a deadly fella.'

There was a filmmaker with us that day too, gathering footage for a documentary about the quarry. I felt strange when Dave asked Carl, not me, whether the filmmaker could borrow a piece of ochre from the site to photograph at home, and then I felt embarrassed about feeling strange. It was a scrap my family was offering Carl and his community. A symbolic scrap on a hill that held no cultural significance for us.

The next day, Ngunawal elder Wally Bell and his son, Justin, fossicked through the rocks scattered across the site. Ochre like this, Wally told me, would've been used in initiation ceremonies at Ginninderra Falls, half a day's walk to the south-west. Pigment washed from a boy's skin into the rushing water below symbolised his transition to manhood.

On the return to the house Wally pointed out the remnants of another scar, largely dislodged by a lightning strike, a few metres off the ground in a gnarled stringybark tree on the ridgeline separating Gollion's east-facing paddocks from those to the west. Wally said it would have been a direction marker (especially at such an obvious spot – on a hill affording a good view all around), and that scars like this were put up high because there would have been more undergrowth here when it was made.

Wally said that for a long time a 'barrier' has prevented non-Indigenous landholders from inviting Aboriginal people onto their land: the belief that in doing so they might lose it. Wally had teased me about this when I suggested he and Justin come inside before visiting the site. '*You're* going to invite us inside for a cuppa?!' I saw it from Wally's perspective: I was wearing my Akubra and R.M. moleskins. And then I saw it from the perspective of a young Carl Brown, running as fast as he could from a farmer on horseback who was probably dressed just like me.

Alighting at our next stop – another scar tree, dead and supine on the banks of Murrumbateman Creek – Matilda gently took my mum's hand. Two Australian women: one a grazier's daughter, educated at an exclusive Melbourne school; the other one of ten children born on a mission in Cowra and sent to the infamous Parramatta Girls Home when she was twelve. If our herd of black Angus stood in the background it'd be the stuff of a corny Meat & Livestock Australia ad.

'Imagine how excited you would have been,' Matilda wondered aloud, 'crowded around this tree as the scar was made.' Dave and Matilda thought the scar, which would have taken hours to cut by skilled hands, was the imprint of a flotation device used for carrying eggs, bulbs or even a baby. Before this creek was turned into a drain by land clearing and overgrazing, it would have linked a chain of ponds supporting birdlife, fish, turtles – and large family groups of Aboriginal people.

Matilda made us all feel the ridges of the tree's scar, the once-sharp cut of a stone axe burnished smooth by a century or more of weather. Her voice faltered.

'That's the part that makes me go all funny. I can't believe it. It's just so magnificent!'

I asked Matilda about her cloak and she immediately became defensive, assuring me she only used feral possums imported from New Zealand. (Thirty-five possums makes one woman's cloak; you need forty-two for a man's.) But now that I mentioned it, did I have any spare kangaroo teeth for a necklace she was making?

There was a kangaroo carcass caught in the fence across the creek, and I squelched through the mud to loot it. By the time I caught up with Matilda (a fox had beaten me to the roo's head), she was atop the ochre quarry with Mum, taking in the view across twenty kilometres of grassy woodland to the slow-motion wind turbines of Ngungara – Lake George.

'What a beautiful place to camp.'

I didn't know if Matilda meant today or 10,000 years ago, and it's this elasticity of time – the realisation that I didn't have to write about the Aboriginality of our farm in the past tense – that made *me* go all funny.

Dave had engendered the same feeling in me during the archaeological survey. He used the conditional tense to describe the hypothetical ('up here [on a hill] is where you'd come to see the comings and goings of your friends and enemies'); and the globalised food system to describe pre-modern subsistence ('the creek would be your Coles or Woolies: here's where you'd come for a feed'). When I asked him if that would've included turtle, he replied, 'Ab-so-lute-ly, buddy. So the mob would've said, "We're sick of eating kangaroo, let's go to Vinnie's and eat some turtle."' *Vinnie's*: I wasn't even in charge of Gollion yet, but Dave was including me in pre-contact Aboriginal society.

On another day, while visiting the quarry with his dad, my five-year-old nephew, Billy, asked Dave, 'Are these cows Aboriginal?'

'That's a good question, young fella,' Dave replied, hands on knees as he crouched to Billy's height. 'These cows aren't Aboriginal because they arrived – or not them exactly but their great-great-great-grandpas and great-great-great-grandmas – with the first people who sailed ships from England to Australia, more than 200 years ago. The people who the English and their cows met when they arrived – *my* great-great-great-grandpas and great-great-great-grandmas – *they* were Aboriginal. As am I.'

Billy looked unconvinced. 'Is this bug Aboriginal?'

It struck me that Dave was able to bring us all together – Akubra-wearing white farmer, Wally, Matilda ('it's World War III between those two, so we better bring them out on separate days'); Carl, my mother and my father (whom Dave called the 'silent enabler)' – precisely because he was No Land's Man. Dave had the professional expertise to gain the government's (and my dad's) respect and the cultural cachet to have the traditional owners take notice, all the while not having his own skin in the game.

The one artefact on Gollion I knew about before his survey was a manuport, a smooth river stone transported from perhaps the Molonglo or Murrumbidgee and used as a kind of anvil to shape blades. Wally had found it, half buried on a rise west of the creek, years earlier with Mum as part of a Landcare walk. When he started digging it out, said Mum, a willy-willy emerged from the creek and whooshed up to them, before vanishing right over the manuport. Wally was spooked, and he carefully re-buried the artefact. Mum says it's one of the eeriest things she's experienced. When I had shown it to Dave and told him the story, he laughed and dug it out with his hands. 'That doesn't scare me, buddy – those spirits don't have power over me, 'cos it's not my country.' This time, no willy-willy formed.

Back at the house, the filmmaker was waiting. Matilda insisted on being interviewed while sitting beside my mum. They were *mudjis* now, Matilda explained. Good friends. As the filmmaker set up, Mum asked Matilda if she'd like a pillow. 'My arse isn't that big!' she replied. Would she at least like to take her insulin in the privacy of my parents' bedroom before the interview? 'I thought I'd just shoot up in front of youse,' joshed Matilda.

While Matilda explained on camera how newborn babies used to be rubbed down with ochre, connecting them to Mother Earth in a way that those born in hospitals no longer are, Mum nodded along like those dorks beside politicians do at staged policy announcements. It was very cute.

'So, you know, these are the ceremonies that people have long forgotten. So that's what ochre means to us: it's special because we can tell our history, our oral history, through ochre.'

'Well, this property here, is, I thought, so very special. When I was introduced to Jane, and her being a good *mudji* now that I've wanted her to be, and she wants to be too, we have learned a lot about each other. And that learning is how we're going to present ourselves in the future. She took me to a place today where the past was ever so present. And seeing that, I had thought, "Well, there's no secret about Aborigines and their past. With wonderful people who belong to this country here with their son, and to know that that will be handed down to another generation of children, living and growing up on this property here, that forever will have the presence of Aboriginal people, in spirit."'

There's no secret, said Matilda, to Aborigines and their past. Nor, I would add, is there anything to fear from listing an Aboriginal Place on a whitefella's place. My family would lose nothing through this process: not the property title, not access to Bald Hill, nor even the right to graze its slopes. Instead we would learn more about our home and form a cultural connection with people who also loved and cared for this country. It's this type of acceptance that Wally had said he was after: the recognition that his people were the first inhabitants of the area.

All parties were in agreement to have the quarry nominated and registered as an Aboriginal Place. Irrespective of the outcome, Wally agreed to give the quarry an Aboriginal name, and my family agreed to allow access to the Ngambri, Ngunawal and Ngunnawal communities whenever they wanted it. It was a hill to us, but in Wally's words, 'the use of the ochre is allowing us to make that reconnection to country.'

It took two years, but in August 2018, the site was gazetted as an Aboriginal Place. And by the time the New South Wales government sent us a plaque for the 'Gollion Ochre Quarry', we were able to ask for a new one – Wally's brother Tyrone had given it the name *Derrawa Dhaura*, meaning 'yellow ground'.

I forget what I was doing when Dave rang and gave me the good news, but I remember his reaction when I asked him if we should hold a party at Derrawa Dhaura with the traditional owners to celebrate.

'Fucken' oath, buddy.'

# 14

It stopped raining. The winter squalls stayed under the doona. The summer storms ran out of juice. Some said it was climate change. They cited the science. Others said it was always so. They cited a poem they had learned in primary school about droughts and flooding rains.

And my father? My father took it personally.

There was no longer much point in watching the nightly news bulletin once the sporting results had been announced and the weather forecast came on, but this was when Dad started paying closest attention. Jolting up off the couch and hulking over the 65-inch flat-screen TV Mum had bought him to watch football on, Dad would accusingly point the remote control at the weatherman and his synoptic chart.

*It'll be another fine and sunny day in the nation's capital …*

'"Fine and sunny"! FINE AND SUNNY! He sounds happy with himself, doesn't he, Sam? Listen to the *glee* in his voice!'

By the autumn of 2019 I was spending more time at Gollion, often staying the night. My fig trees were starting to resemble an orchard (there were now eighty trees; I picked my first fruit the previous March, though only enough for me to eat on my muesli), and to minimise evaporation I turned on their drippers at dawn for four hours each week.

One morning I passed my father in the kitchen at 6.45 a.m. He was wearing nothing but chequered boxer shorts and was holding a transistor radio to his ear, jugular veins bulging, the early morning light catching the silver neck hairs he'd missed shaving the day before. He shouted down the hall to my mother, who was propped up on an elbow in bed.

'Ahhhh SHIT!'

'*What?*'

'The airport got five mills!'

'*I told you we were on the edge of the rain band.*'

'We're always on the edge ...'

'*No we're not, David ...*'

'I'm just so sick of this ...'

The night before, when I had asked if he'd like a wedge of rockmelon for dessert, he replied, 'I don't care. I don't care about much anymore these days.' It was then Mum told me Dad had self-diagnosed himself with depression; *The Land* had published a questionnaire for primary producers, right where the crossword used to be.

Gollion had known drought before. Farmers in the district still talked about 1982. That year, people shot their animals in the paddock and left them for the crows. It started raining again when my family moved in. The 'Millennium Drought', a long period of lower-than-average rainfall, lasted the first decade of the twenty-first century and prompted Dad to stop spreading superphosphate ('to give the native perennials a chance') and to start rehabilitating the creek ('the best place to store water is in the ground'). One hundred years earlier, the 'Federation Drought' dried up the Murray River and forced thousands of small farmers to walk off their land. Fernleigh's well was the last in the region to run dry, and farmers throughout the valley queued up their buggies to replenish water tanks.

But nobody alive had seen anything like this. Gollion's annual rainfall is 630 millimetres. A plastic rain gauge on the swimming pool fence was eagerly checked whenever it rained and, as per meteorological convention, its contents at 9 a.m. added to a chart on the kitchen wall as that day's rainfall, even if it was still raining. The recorded rainfall for 2017 was 583.5 millimetres. In 2018, it was 475 millimetres. In 2019, just 292.5 millimetres were emptied from the gauge.

In August 2018, in Europe, much of that continent suffering its own severe drought, the residents of Děčín, a historically German-speaking town in the Czech Republic, were granted a harbinger of hard times. As the water level of the Elbe River, which passes through Děčín, fell, it exposed a series of 'hunger stones', boulders chiselled with the years of droughts past – and warning of their consequences. One stone contained the following inscription:

'*Wenn du mich siehst, dann weine*': 'If you see me, weep.'

At Gollion, too, the dry weather brought a sudden, morbid clarity to the landscape. Tools showed up where no grass could hide them. (Though unfortunately not the lost crowbar Dad had given me for Christmas.) Spools of wire, left over or forgotten from long-ago completed fencing jobs, now suddenly easy to see, were chucked on the back of the farm ute before they could tangle its driveshaft.

Too-well-hidden Easter eggs glinted in the sunshine, their stale chocolate tossed onto the compost heap.

Fernleigh's dams fell to the lowest level since my parents incorporated the property with their own in 1994. Some had clearly been used as a tip by the previous owners (for car tyres and offcuts of corrugated iron). Across the entire farm, our efforts to cover the soil with native perennial grasses were exposed as sorely wanting, with whole hillsides unearthed as annual grasses shrivelled up and blew away.

The country was revealing its secrets for the second time in two years.

We were regenerative farmers, Dad said. We cared for the country. But as the drought progressed, our actions suggested otherwise. Grazing school had taught me to match the stocking rate to the carrying capacity; in the natural ecosystems we were trying to mimic, during dry spells herds move on from a bare grassland – or die of starvation. This was consistent with my father's economic ideology. The best way to cope with a drought, he said, was 'money in the bank': sell livestock, don't buy in hay at inflated cost ('as in wars, scoundrels always make money out of droughts'), wait for it to rain. Then, once it does rain, either buy in heifers that are ready to be joined with your bulls ('get rid of your bulls last'), or wait. The high quality of Angus cattle in Australia, said Dad, is such that it's no big deal to restock with someone else's cattle. It can be expensive to repeatedly pay a truck driver to take small loads, but sometimes that couldn't be helped.

At grazing school, Brian had taught us to switch the emphasis on what we farmed from cattle to grass. The former can be bought and sold with a phone call; the latter can take decades to recover from mismanagement. In times of drought, he said, your decisions should be made in the interest of the landscape. And even in non-drought conditions, the best holistic managers reserve a paddock of spring grass as a 'standing haystack' in case a drought develops. By then, the dry feed won't be as nutritious as if it had been grazed in spring, but it is still preferable to buying in hay. Changing this paradigm, from farming cattle to farming grass, is sometimes called 'changing the paddock between your ears'.

Conventionally, droughts in Australia have been managed for the health of the livestock, not the pastures. Graziers would put their sheep or cattle into one or more 'sacrifice' paddocks, where they would remain until it rained and be fed grain and/or hay. The overall health of the paddock would be sacrificed to the animals; not a blade of grass would be spared. You can pick out old sacrifice paddocks driving through the countryside: in spring, they sprout fast-growing annuals first, as they have accumulated a high concentration of nitrogen in the form of manure.

Making or buying in hay was discouraged at grazing school, for financial and environmental reasons. Holistic managers are critical of cutting hay because it enables farmers to stock landscapes that are in need of rest, and because it disrupts the ecological process of nutrient cycling. If a paddock is cut for hay and that hay is transported elsewhere, the farmer is not allowing the nutrients in those plants to return to the soil, either as dung or as trampled plants.

Dad liked to have some hay in the shed, for 'insurance if things get really bad', but Brian taught us you should have destocked by that point. That 'point' usually comes in autumn, as this is the last time before spring that the soil is warm enough to explode with grass if rain falls; a dry autumn will prolong the drought until at least spring, as any rain that falls in the coldest months won't create significant grass until September.

After 2018 brought the third dry autumn in a row, we started reducing the size of our herd, beginning with the usual weaning of calves born the previous winter 'onto the truck'. On 5 April we consigned forty-eight calves to a store sale, with a plan to sell older stock at a fat sale if conditions didn't improve soon.

I had a habit of daydreaming in the cattle yards. My job was usually to open the drafting gates while Dad used his judgement to separate the cows we wanted to keep from those we didn't. He used a piece of poly-pipe to prod the ones he wanted and let the ones he didn't pass him, where I would let them out before closing the gate. Sometimes a beast he wanted inside the yards would make a dash for the gate, which is when I would have to be paying attention. I wasn't always paying attention. Something about the dappled shade of the yellow box tree above, the lowing of cattle and the way the sun hit the dust particles – particles increasingly so light and dry they could have been cocoa powder – made me dopey and slow to react.

Weaning this way is normally traumatising for mother and calf. The 'weaners' would cry all the way to market, while their confused mothers would bellow through the night, a sleep-interrupting grief that could last up to a week.

But that day, as I helped Dad muster, he pointed out that the older cows 'couldn't give a toss' about being separated from their calves. (The cattle truck had just left but was still audible, clattering down the lane, wheels spinning up plumes of dust.) I had a look and could see what he meant. Some of the younger mothers had already found their calves (both the ones deemed too small to go to the sale, and those spared the journey because there wasn't enough room on the truck). The older cows were silent. Dad thought this had to do with the drought – maybe it's a relief no longer to have a mouth to suckle when there's barely any grass to eat.

We were back in the yards in May, no rain having fallen since the previous sale. This time we selected our fifteen worst stock, to be sold at a fat sale. Some were dark brown cows, throwbacks to a time when there were non-black bulls on the property. They were sold to 'keep our line black'. Some were sickly looking steers that had failed to fill out (Papa, said Dad, called such steers 'hollow'). In marginal country like ours, Dad continued, it was wise to be slightly understocked and to carry low-risk stock like steers over cows. In a drought, steers will be easier to sell.

Once that truck was clattering down the lane, we still had 130 head of cattle and a diminishing amount of grass. And so we started feeding them hay.

Haymaking: it was a highlight of my otherwise farming-uninterested childhood. I loved every step of the process, from pressing my face to the school bus window on the way out of Gollion past the paddocks Hay Lower and Hay

Upper as the massive lawnmower left rows of clippings where once there had been grass; the 'tedding' that would be the next step (another implement is attached to the tractor to fluff up the clippings to help dry them out); the raking a few days later into lines ready for the last process, when a massive baler would vacuum up the rakings and spit out tightly bound rectangular bales, to be picked up like Lego off the carpet and stacked into sheds from the floor to the ceiling.

Dad had last made hay in 2006, feeding it out in a few dry spells since, and on exceptionally cold days, or when a calving mother had to spend the night in the yards. Hay has a shed life of twenty years, but by now it was old, stale and not very nutritious.

I'd helped Dad to roll out hay bales before and liked the feeling of them gaining momentum, unravelling as they did, becoming easier to push with each turn. I remembered he would stamp on the mice that emerged from underneath bales as we rolled them out of the shed, a commitment to pest control I found both funny and extreme.

One morning that June of 2018 I drove us to the next paddock in the cattle's circuit and we unrolled the bales for the herd to eat later that day, first cutting the netting with a blunt Stanley knife Dad kept in the glovebox, then rolling each one out until a Japanese fan of hay lay waiting in the dirt. On the way back to the house I noticed a lamb's head sticking out of the back of its mum. We put all the sheep in the yard, and I held the ewe while dad pulled out the lamb, which was breeched. At first I thought it was dead, but then it started blinking and, with the encouraging lick of its mum, was soon taking its first tentative steps.

There were other bright spots amid the gloom. I went to a regenerative agriculture conference in Wodonga with my parents, where Dad had been asked to speak about his creek restoration work. He got a standing ovation. On the way back, I asked what he liked most about farming. Growing fruit, he said. When I asked him why, he said he liked being able to share it around.

A few weeks earlier, while fencing off more of our creek restoration work, Dad had turned to me unprovoked and said he thought this project was the best thing he'd done on the farm. That had been a great day. It was cold, but because we were fencing, I was soon only wearing a thermal top, sleeves rolled up to my elbows. It was foggy when we arrived at the creek, and when the fog cleared it revealed a full moon that was visible until late morning. We put in new fenceposts and then strained up new wires. First I tapped in the steelies with a hammer; then we took turns to bang them in with the petrol-engine pile-driver that Dad had once hidden from me. It was even heavier than the manual donger, and every ten or so posts (he steadied them while I drove them in), Dad would say, 'Have a spell', and take the driver himself.

The winter of 2018 ended without rain. One day Dad said he thought 'the way for me to go' would be to run half as many cattle as we did (that is, about sixty-five instead of 130) and in a good autumn sell calves at 300 kilograms; in a bad year, at 200 kilograms. With a lower stocking rate I would be less likely to run into environmental or financial problems. He was acknowledging that we were now in trouble.

Spring came and with it more mouths to feed. Between mid-August and early September, new calves arrived daily. 'That one was born this morning,' said Dad of one lanky little fellow, fur slick where its mum's spittle had dried. 'You can tell by looking at the mum's bum.' (He was referring to the trail of afterbirth hanging out of its vagina, a frozen waterfall of purple slime.) Dad didn't say it, but I'm sure we both thought it: you could make a killing flogging this stuff to the paleo cafes in Sydney.

The new year brought no new rain. Facing the prospect of another dry autumn, on 31 January 2019 we drafted thirteen cows to sell (the number that would fit on the truck), selecting those least likely to carry a calf through the winter. By now there were fat sales in Yass every week. Because we'd previously separated most cows from their calves, it wasn't as easy to spot the dry cows, as the mothers who'd had their calves taken off them in the spring were beginning to look dry themselves.

Increasingly desperate for ideas, we turned to another input: molasses. Cows love it, Dad said, because it feeds the bacteria in their rumens, allowing them to eat exceptionally dry fodder. We opened a drum and poured it, thick and treacly, into an ice-cream container and from that onto tussocks, thistles, wild rose bushes – plants the cattle wouldn't normally even consider eating. Without the aid of molasses, Dad said (or a salt lick, which performed the same function), cattle couldn't fully digest the driest feed and would begin to lose condition.

'With molasses you can feed cattle nothing but old hay – maybe even wood chips.'

What did feeding cattle molasses, I wondered aloud, do to the amount of methane they produced? And *wood chips*?

'Molasses is lovely stuff,' Dad continued, oblivious to my questions. 'My father used to put it on his Weet-Bix.'

'Didn't your father lose all his teeth long before I was born?'

'Yeah, but he ended up living until he was ninety-three!'

The livestock agent rang and told Dad one of the thirteen cows we sold had fetched over $1000, the highest price at the Yass saleyard that fat sale. 'Not bad, eh,' said Dad, relaying the news in surprise. 'Yass Queen!' I said, but he wasn't in the mood for joking anymore.

In March 2019, the start of the fourth autumn in succession with no rainfall, we sold our fifty-four biggest weaners and then fed the thirty-three that remained a diet of calf nuts (a kind of vegetarian kibble for calves) and grass-clippings (carefully raked up from my fig orchard) for a few weeks, before selling them too. Each weighed only 150 kilograms when we put them on the truck (in a non-drought year they would be at least 250 kilograms).

Then, on the last day of March, we received 41 millimetres of rain. It was an immediate reprieve, but if we didn't receive a follow-up by Anzac Day, said Dad, the 'green pick' of grass it prompted to grow would wither, and we would have to reassess once more. We received no follow-up rain by Anzac Day. 'What's the point?' Dad said, by which he meant continuing to keep cattle. I didn't have a reply.

We were now reassessing our options on a weekly basis.

On 3 May, 30 millimetres of rain fell. Dad thought it was enough to hold off selling more stock just yet. Fifty-four cows and three bulls remained. How would we get them through winter alive?

For two years now we had been waiting for a drought-breaking rain. It was a form of gambling, and the house was stacked against us. My father, so used to being capable and in control, was now lost and helpless. He worried constantly for the livestock. Mum worried for Dad. I worried for the land. And in *The Land*, among the burgeoning pages of cattle sale results, were articles about dealing with depression. Mum cut out an ad for an anger management course and left it where Dad would find it, on the kitchen bench, when he came inside after fruitlessly checking the rain gauge.

One thing I didn't worry about that year was getting booked on the drive to and from Sydney. My little car could run the gauntlet of speed cameras in under three hours, but for an extra fifteen minutes (provided you boarded in time to avoid sitting within smellshot of its flapping toilet door), you could sleep on the Boyfriend Express and let its driver keep an eye on the speedo. Southbound, we called it the Girlfriend Special.

Her name was Lauren. When people ask if we met online it is fun to confirm that we did, then watch their confused faces decide which is worse when we clarify the platform was Twitter not Tinder.

Initially we messaged about movies and writing. She was a film critic and essayist. I was spending every non-farming day writing about a murder trial. We met for a coffee in Redfern that turned into several beers. Her vocabulary – 'normcore', 'identity politics', 'late capitalism' – was not that of Gollion's father–son farming sessions. At one point during that first meeting an enormous rat squeezed out of the downpipe beside us, shook itself dry and galloped off in the direction of Surry Hills. Cities, said Lauren, are unfit for human habitation. I thought: maybe this could work.

I was then still living in Canberra and we fell into a routine: alternating weekly, Friday nights one of us would board the Boyfriend Express/Girlfriend Special for the other's city. Sunday afternoons we would reluctantly part, already planning the next weekend. In between we spoke on the phone every night: me from the stoop of my share house or, increasingly, the driveway paddock of Gollion, where reception was attainable if you stood between a holm oak and a white cedar, and then only if you held your body straighter than either tree; her from the attic of the terrace she rented from an octogenarian widow who lived on the first floor with her two overfed cats. It was nice having a man around, the widow said, and each time I visited she would ask me to do manly things like empty the lint catcher or change a lightbulb.

The first time I took Lauren to Gollion we walked to the creek. I was preparing to give her the Reed Warbler spiel when we arrived at its bank and instead found the neighbour's sheep. They had found a hole in the fence (the neighbour's pastures were even barer than ours) but couldn't fit back through it by themselves. I tackled each one and dragged it under the fence, but the last was too big for one person and was panting with exhaustion. Sheep, I remembered Dad telling me, 'have a habit of dropping dead on you'. I asked Lauren to help me lift it over the fence, and she gamely agreed. Watching her wonder where to grab hold of the thing made me realise the farmer I had become.

For most of that first year, her main contact with Gollion came each Thursday morning at 10.05, after the news headlines had been read and her co-host told listeners it was time to 'talk streaming' with the national broadcaster's television critic, my girlfriend Lauren. Sometimes I propped the transistor radio on top of a ladder as I patched a hole in the orchard netting; sometimes I kept it in my breast pocket. If I was out in the paddock, I made sure I took Dad's 'good' ute – the one he claimed for taxation purposes despite it barely leaving the road. It was the only one of Gollion's farm vehicles with a still-working radio.

I have never spent so much time in a big city as I did in Sydney that year. I would have had fun anywhere; we were in love. We slept in late in Lauren's pitch-ceilinged bedroom then ate luxurious breakfasts out and walked through the Botanic Gardens, stopping to read books beside the weeping mulberry. I liked the way the city was built into the environment, laneways ending suddenly in sandstone dead-ends, trickles of water and vines running down cliff faces. But some things I never got used to. At dusk in Lauren's neighbourhood the food delivery bicycles teetered to work, their faint lights flickering like so many drunken fireflies. Who were their customers, putting these barely upright cyclists in peril? But self-catering brought its own reminders of a food system gone bung. At Lauren's local supermarket, in one of the country's most expensive postcodes, the only organic produce looked to be lying in state.

What would become of us? Lauren proudly subscribed to the cultural cringe, was unsentimental about her hometown and planned on living overseas as soon as she could. My deepening love for Gollion was now so blind and unquestioning I had come to defend even the *powerlines* when visitors commented on what a shame it was that they 'blighted' (I preferred 'dotted') the property. Without addressing us specifically, Lauren talked about the compromises all couples make to stay together. I nodded, but couldn't see what room for compromise there was in my plan to take over Gollion.

Was it a form of arrested development I had, this desire to continue my parents' legacy? When Lauren stayed at the farm, we slept in the bedroom I'd used as a boy, glow-in-the-dark astronauts moon-walking on the ceiling. If this house were in the suburbs, such sentimentality would be weird, grounds for a break-up: more proof, if any were needed, that my generation cannot grow up. I could never expect Lauren to love Gollion the way I did.

Two years later I would hear a farm succession consultant, Lynne Sykes, speak at a conference. She said she once heard someone describe her as 'the big sheila from Dubbo who's always banging on about divorce.' She is. She does.

Sykes describes seeing many women leave farming marriages *after* a drought breaks. She doesn't specify why these women don't leave their marriages during the drought, but I figured the implication was that an ex-husband would be preferable to a suicidal one. Once the rains return and mental health, like the roadside fire danger warnings, have subsided from catastrophic to their usual extreme, women feel they can safely leave. The dry spell has made it clear to them what is most important to their husbands, and it is their farm, not their wife. (Lynn: 'My husband worked in agriculture, and all I know about farmers is that they had no respect for family time. They would ring at 10 p.m. and

6 a.m., and I thought, "Gee, don't you people have a life?"')

Would I have to choose between living on the farm alone or living nearer a metropolis with Lauren? When I told my best friend, himself a farmer, that I was in love with Lauren but she was not equally taken with Gollion, he replied as if he'd heard it all before: it was a quandary, he said, as old as farming itself.

I figured there were still a few years until I would take over Gollion, and in a few years, much could change – I was still clinging to Lauren's rat-provoked comment as a sign she might be ready for a tree-change. But then one day, after lunch on 31 July 2019, Dad looked up from the jerry can we were emptying into an old Pajero we used as a farm vehicle and said, 'I've had a gutful of this drought. I think it's time you took the plunge and moved to the farm.'

# 15

What would Gollion be like without my father? Or: was I ready? There was not yet a timeframe for my parents' departure. Mum told me they wanted to wait until the drought broke and Lauren moved in, which I took as a subtle reminder that Gollion was a team effort, as well as naively presumptuous on both counts. Dad stressed that when they did eventually move to the coast, it was only two hours' drive away, and he was 'only too happy' to help with calf-marking or orchard-pruning when required. I noted the enthusiasm in his voice, despite his self-diagnosed drought depression. It raised a further question: what would my father be like without Gollion?

By moving full-time to the farm in the spring of 2019, I felt I had entered the final stage of the succession plan. Gollion was now my home as well as my workplace, with no clear distinction between them. A pregnant cow goes down with a breeched calf just as you're sitting down to dinner? Dinner can wait. On your way to the movies, you notice that the neighbour's bull has pushed down your boundary fence? You're not going to the movies anymore.

I was still commuting to my office job on Mondays and

Tuesdays, but every other weekday, I farmed. Sometimes I worked alongside my father; sometimes I worked alone. There was no longer a schedule, or tasks Dad saved up for us to do together when I was there. I couldn't yet do all the work of a farmer (I didn't have my gun licence, an unavoidable necessity for euthanaising, and most things mechanical I found confounding; I could often be found holding a spanner in one hand and scrolling YouTube how-to videos with the other), but by now I knew what the work was, and that it fell into two categories: jobs you could put off until tomorrow and jobs that needed to be done now.

I had no way of knowing if I was ready for the takeover, and in the way of these things – getting your driver's licence, becoming a parent – I suspected I wouldn't feel ready until I was. Whether my father thought me ready weighed more heavily on my mind. That September, on one of my first nights back in my childhood bedroom as a permanent resident, I dreamt that I drove my little hatchback up Gollion's driveway, parked by the compost heap and turned off my engine. But when I opened the car door, I wasn't met by Suey Dog pawing at my lap or Dad pushing a wheelbarrow, but a whiteboard and an easel. On it, in blue permanent marker, my father had written:

'Clearing sale starts 3 p.m.'

A clearing sale is an on-farm auction of machinery, tools and any other paddock-strewn junk the real-estate agent has advised the farmer to flog so they can tidy the joint up for the property supplement of *The Land*.

I woke myself up before any buyers arrived: this, then, was my nightmare. I had failed as Gollion's custodian and now my parents were selling up.

Waking hours held clues that Dad, too, was unsure. One day I helped him hang a new gate in the corner of Dead Horse paddock. He was wearing a hemp hat my Uncle Michael had given him as seventieth birthday present. The brim was not especially wide, and a wire running through it allowed the wearer to adjust its shape. My father had adjusted it so that the brim frilled around his face like a giant clam shell.

We cut the existing fence, sunk a new hardwood post, strained the existing wires to the post and hung a gate. We'd done this several times before, but Dad reiterated what to remember as we completed each task: when drilling into hardwood, make the pilot hole slightly smaller than the screw that will follow it; only use 'coach' screws, which are designed to fix metal to timber; when positioning the lug on the post that the gate will hinge on, put it high enough to ensure the gate swings easily but low enough that a bull won't be able to stick its snout underneath and lift the gate off the lug.

Dad stood back, removed the drill bit from the drill and threw it with a clank into one of his mother's old cake tins.

'This gate will save half an hour each rotation.'

By this he meant the farm-wide rotation of the herd. The gate we had hung created a shortcut across the lane, from the Gollion side of the farm to Fernleigh. We were ready to test the new shortcut, but with the sun already hot by 10 a.m. I suggested we first return to the house; Mum was in Sydney and wanted us to water her veggie garden before the hottest part of the day. The patch was a new one, in a fenced-off section of the chook shed, watered by hand with a bucket. The plants – tomatoes, zucchinis, capsicums, chillies – were badly wilted.

Dad started ranting about why there was no hose to water them.

'Mum should be able to grow all her veggies, but she still goes to the farmers' market every Saturday. I mean, she grows her veggies to get a buzz out of it – like I get a buzz out of growing my fruit – but I dunno why she persists with these little-old-lady systems!'

In his frilly hemp hat, I said, he looked like a little old lady himself. I told him I was going to turn off my fig drippers and would meet him back at the house for a coffee. Now it was my turn for a spray.

'You spend a lot of time mucking around with your figs. That's what I worry about you ...' He didn't finish the sentence; there was no need: I was inefficient, like my mother. I never would have thought of hanging that new gate. And he was right.

For Dad, the trained economist, farming – life – was about efficiency, about increasing productivity. I was a potterer. When I noticed that some oak trees I'd recently planted in the lane needed watering, I made two trips, carrying buckets from the nearest tap, instead of one trip in a ute or on a motorbike with a larger water container. I didn't consider it a waste of time to walk in contemplative silence up and down, and then back up and back down, the driveway paddock, breeze on my face, galahs passing overhead – I considered that life.

I didn't mind that we were different. I wouldn't be the same farmer as he'd been, and I felt I didn't need to be as efficient: I would have more time to do farm work, because (eventually, I hoped) I wouldn't be juggling the farm with an office job in the city like he had for so many years. Maybe

I was delusional, but I wanted to make up for that loss of income by being even more self-sufficient in what I ate than my parents had been.

There were other ways we were different. I was more patient than Dad. And more optimistic. Dad's attitude to farming, like his attitude to life, was to be pleasantly surprised when things worked out the way he wanted. At three-quarter time in the 1993 AFL preliminary final, one of the first football games I remember watching with him, his beloved Essendon trailing Adelaide by just two goals, he pulled on his hat and declared, 'It's aaaaallll over', then disappeared to do some farm job. He returned in time to see the victorious Dons singing the team song. That catchcry – *it's aaaaallll over* – has been used by my family ever since when a minor obstacle presents itself.

But the drought was making us all pessimistic, impatient. Blunt blades frustrated Dad the most. He was always complaining that Mum 'never has any sharp scissors' – or, worse, she did but hid them from him. ('I've seen them – I know she has some.') One day that spring I was cutting a piece of Geotech fabric for a vegetable bed with some scissors I'd found in the kitchen. It was slow going, and Dad offered to take over. He lasted less than a minute.

'God, I *HATE* Mum's scissors. I'm *too* impatient and *too* old to use blunt scissors!'

He'd long admitted to impatience, but this accompaniment was a new one – that he was old. He still had his farm strength, but I had begun to notice some mental slippage. Mum said she was getting more forgetful, but she'd been leaving half-drunk cups of tea around the house my whole life. Dad's decline was more evident.

He would do things like leave one of the farm utes' doors open overnight, causing the battery to go flat. He was also becoming deaf. He'd eat his lunch and leave the dirty plate on the table, without even pushing his chair back in. He used to at least put his plate in the dishwasher. It seemed unfathomable that Dad – my dad, the do-it-yourselfer 'fifties boy' – would ever lose his faculties.

On 3 November we had 18 millimetres of rain ('enough to give us four weeks to play with'). I'd been at work in the city and met Dad at the cattle yards on my way home, where he was fiddling with a water trough. I told him I was going straight to the rain gauge – I wanted to see the proof for myself.

He replied: 'Going to a Landcare meeting, are you?'

Another day he drove us to the creek. He wore a striped shirt I'd not seen before – at least not before bedtime.

'Are you wearing pyjamas?'

'Yeah. Why?'

'No reason.'

With age came sentimentality. On my birthday that September my parents gave me a card with a photo of a dainty little fairy wren on the cover, chosen, I am sure, by my mother. Inside, she had written:

*Happy Birthday Sam,*

*With much love from*

*Mum, Dad*

Mum had written that she was enjoying 'pottering' with me in the garden before handing the pen to Dad, whom I imagined scratching his head as though I were a retiring work colleague whose name nobody could quite recall. *Dan*? *Cam*? But in his messy scrawl, my father had added:

*I am enjoying working with you on the farm*

For a man who shook his son's hand on the eve of a long departure instead of hugging him goodbye, this was soul-baring at its most raw. I still keep that card on my desk.

By now the orchards were the only source of greenery on the farm, and they were green only as far as the irrigation drippers could reach. One afternoon we walked together through my fig orchard, removing the *breba* (first) crop of fruit while they were still small, to encourage each tree to put its energy into the second, better crop. The trees were growing quickly and despite the drought I hoped to have the first saleable crop from my new trees the following autumn.

We identified grasses together as we walked: Yorkshire fog, barley grass, cape weed, cocksfoot (he ran his fingers through its blades to check for identifying serrations). It was a drought, but here at least it was also spring, and it was nice to pretend that everything was normal, with flowers and leaves and the sound of baby magpies nagging for food.

We paused for a moment: me on my haunches, Dad lying down in the manner of a come-hither kangaroo. I asked if he wanted to come inside for a cup of tea.

'I'm just going to keep lying here a bit longer. It brings me so much gratitude, this orchard.'

There was now an urgency to our farming tutorials that went beyond the desperation of the drought. It was as if a pop quiz could be called at any moment, and we crammed the remaining curriculum into the next few months. My notebooks from that spring are full of new lessons – grafting, pruning (refresher course), imprinting, plumbing (refresher course), welding, driving a tractor, 'dropping' a tree, drilling into steel; pump maintenance, chainsaw maintenance, motorbike maintenance, spray-unit maintenance – with accompanying safety tips to avoid following Dad's well-trodden path to hospital: 'Never weld a fuel tank, even an empty one. Fuel fumes stick around for a long time'; 'Don't use Google Images to identify plants'; 'When you have forks attached to the front and back of the tractor, it's suddenly very long. Keep this in mind when turning through gates'.

We prioritised tasks that would be hard for me to do on my own. Dad used to talk of drought-proofing Gollion (by building more dams, slowing down the creek's flow, storing more water in the landscape, subdividing more paddocks and planting more trees). He hadn't mentioned drought-proofing for a while – droughts, clearly, could not be avoided, only managed. But it seemed like he was now Sam-proofing Gollion. He would leave it in as good a shape as he could, and no matter how badly I fucked up, at least the gazebo (home-made, like the tennis court – my father's only two concessions to the pretensions of the bunyip aristocracy) now had a new skylight, and the orchards were now fed by a new solar pump.

During a heatwave in November we put down new floorboards in Fernleigh's hundred-year-old woolshed and replaced its rusted sheets of corrugated iron with shining new ones, too hot to handle without gloves. It was regrettable, said Dad, that he'd allowed Fernleigh's other historic sheds to crumble on his watch, but at least this one would now stand into the future.

'I owe them that much, at least,' he said.

'Owe who?'

'The early settlers.'

A cold Saturday in July. Fast-moving clouds, low and dark and carrying their moisture to some lucky buggers elsewhere. At midday they appeared: a single-file line of guests walking the two kilometres from the carpark by Fernleigh's woolshed up into the hills. For those who couldn't manage the walk, two utes, driven by Dad and a friend, ferried them up and back.

*Bald* Hill never did have much coverage up top, but by the time we got around to decolonising it slopes it had gone full Sinéad O'Connor. The grass was green from the dew of winter, but the drought had kept it to the length of a matchstick, what farmers call 'pick'. When I later saw drone footage of the hill that day I noticed large chunks of it had no grass at all, overgrazed back to the ironstone beneath, itself flecked with veins of yellow ochre – of derrawa dhaura.

The morning of the ceremony, Wally took me through it again, reminding me to roll the Rs and emphasise the second-last syllable.

Derrawa Dhaura.

*De-ra-wa Dhaaaauuuuur-a*.

I watched how he made the sounds, the lump of his tongue pressed against the bottom of his mouth, his white goatee jutting outwards. I wanted to pronounce it 'in language' – I wanted to pronounce it right.

We'd capped the event at 100 guests, but who knows

how many more came? Ngunawal mob. Ngunnawal mob. Ngambri mob. My family and our friends. Randoms: friends of friends; a New Yorker; an Irishwoman in high heels (a neighbour we had never met, she had heard about the event on the radio that morning).

*The Canberra Times* sent a reporter and the ABC a camera crew. Fires burned to keep guests warm and caterers served food. Kids rolled down Derrawa Dhaura's eastern slope while on its rocky summit, Dave Johnston led archaeological tours. Guests – *folks* – hung off his every word.

There were speeches, each speaker introduced by Dave: Wally first, then Mum. I can't remember much about either, because I was nervously waiting my turn.

Here is what I said:

Thanks, Dave, and thank you all for coming.

I'd like to acknowledge the traditional custodians of this country.

The Ngambri people.

The Ngunawal people.

The Ngunnawal people.

I acknowledge two present elders here today, who have both been instrumental in the protection of this site as an Aboriginal Place. Wally Bell of the Ngunawal people and Ngambri woman Aunty Matilda House.

And, with the permission of his family, I acknowledge Carl Brown, a past Ngunnawal elder who visited this site in 2016, and who was proud to support its listing as an Aboriginal Place.

One night, two weeks ago, I received a text message from my mum. It read:

'Wally says Derrawa Dhaura. It means yellow ground.'

I'm happy to admit that I teared up when I read that. And not only because it means we can stop calling this paddock by the incredibly boring name Bald Hill.

I teared up because in giving this yellow ochre quarry a Ngunawal name, Wally Bell was giving it new life. And I thank you for that, Wally.

This place has been used by Ngunawal, Ngunnawal and Ngambri people for thousands of years; it was part of the regional network of ochre quarries that includes the red ochre quarry Gubur Dhaura, in what is now the Gungahlin suburb of Franklin. Just over those hills.

But by the time my family moved to this farm in 1983, it had been a while since yellow ochre was collected here.

My sister Eve has written of the sense of absence she felt in this landscape as she walked it as a teenager, and the older I get, and the more time I spend working here, it is impossible for me to ignore what that absence means.

Aboriginal people had not collected ochre from this country for a while when my family moved here, but not a long while. When the village of Sutton celebrated its 150th anniversary in 2017, I came across an article from *The Canberra Times*, written in 1979, about the district's first European settlers. In it, Cecil Thompson, a well-known member of the Hall community, recalled his great-grandmother telling him, when he was a little boy and she an old woman, of the huge corroborees she saw growing up at Tallagandra, just over those hills.

That history is within touching distance. Some of you would've met Cecil Thompson.

'Another farmer's wife,' Thompson told the *Canberra Times* reporter, 'was also scared of the Aboriginals. But her husband used to take care of the problem whenever the tribe appeared on their land.

'He used to call the chief in, hand him some tobacco and tell him "Get your mob to move off."'

That's not good enough.

When Dave walked this slope in early 2016 and realised what it was, he invited Wally, Matilda and Carl to walk it with us. In learning the archaeology of Derrawa Dhaura from Dave, and its cultural significance from Matilda, Wally and Carl, my appreciation for this country – Ngunawal Country, Ngunnawal Country, Ngambri Country – was deepened greatly.

When Carl Brown came here, he told me the last time he'd been on a white farmer's land he was a little boy, near Yass, hunting rabbits, and that this white farmer chased this little black boy off the farm, on horseback.

That's not good enough.

As non-Aboriginal landholders, we have an opportunity to take meaningful, grassroots action in what I think, at a national level, has become a stalled project.

Truth-telling. Reconciliation.

Two Aboriginal Places in New South Wales on private property were gazetted in 2018: Derrawa Dhaura and Millpost's axe quarry.

Both are on Ngunawal Country. Ngunnawal Country. Ngambri Country.

But the reality is, there are thousands of such sites, across this continent, that don't have the good fortune to be located on farms owned by mates of Dave Johnston.

I hope the listing of these two sites as Aboriginal Places, and all the good that has come of it for all parties involved, prompts white farmers elsewhere to not be scared, but to be inspired.

But nor do I want this Aboriginal Place listing to be superficial or symbolic.

In giving Derrawa Dhaura a name, Wally has indicated that the Aboriginal cultural significance of this yellow ochre quarry does not belong to the past.

Me and my family are here to facilitate your return to this place, Wally, and you, Matilda.

And your children, and their children.

How you use it is up to you. It's yours.

Thank you all for coming.

Photos were taken. A plaque was unveiled – fittingly, I thought – by Mum and Matilda, each woman wrapped in a scarf against the cold. In the three years since first meeting they had become even better *mudjis*, and the chest freezer on Gollion's veranda was full of kangaroo tails shot by my brother-in-law Shane and awaiting delivery by Mum to Matilda.

On the TV news that night, I was surprised how magnanimous the elders were. In his speech, Wally emphasised, as he had on the short film made in 2016, that property owners in particular 'will not lose their place because of these significant areas and places within their properties. We only want people to learn about our culture and how we practise our culture on country.'

Matilda was sitting down in her interview. She was tired. The end of a big day. 'I'm very happy about what's happened.

Yeah. Long time coming. But we need more onboard to let them know that we don't want your *land;* we just want the history that's there, to share it with everyone else.'

Regenerative agriculture, relatively new and as yet uncommodified, has no single definition. Joel Salatin, perhaps its most famous proponent, defines the practice as farming that grows food while improving the commons: soil, water, air. Other definitions encompass financial as well as ecological regeneration and, more importantly, social regeneration. Environmental historian George Main defines regenerative agriculture in relation to history. Colonisation and industrial agriculture, he argues, shattered ecological systems and Indigenous societies. In this telling, regeneration refers not just to healing land, but also to healing people. The cultural renaissance of Derrawa Dhaura is inseparable from the ecological recovery of the landscape that hosts it.

How many Australian farmers would share this view? Dave had told me several anecdotes of priceless artefacts hidden or destroyed by farmers, afraid of the consequences, real and imagined (mostly imagined), of what would happen if they declared them.

People who are comfortable with the foundations of their society do not behave this way.

# 16

There isn't a vocabulary for what happened next – summer bushfires that began in winter; millions of acres burned and *billions* of animals killed; temperature records set, only to be broken – but faced with the indescribable, Australians made do with the words we had.

The Australian summer of 2019/20 has come to carry a prefix previously applied to individual days. There had been a Black Tuesday and a Black Thursday; a Black Friday and, most recently, Black Saturday. Now there was the Black Summer.

On 12 November, Sydney was placed on 'catastrophic' fire danger for the first time since a new rating system was enacted after the Black Saturday bushfires of 2009. The day before, no rain fell on the Australian continent for the first time since rain gauges were installed. It was, said *The Sydney Morning Herald*, 'the day it forgot to rain.'

I was leaving the house for my office job on the morning of 12 November when I ran into Dad. He had been walking around the farm, assessing whether it felt like a bad fire day. He thought it didn't, because there was still a green tinge to

Gollion's bare hills, and although it was windy, the forecast maximum was only 28°C. 'I've been in a lot of bushfires,' he said. 'And it doesn't feel like that.' Turns out he misheard the radio – he didn't realise the 'catastrophic' rating didn't apply to the Southern Tablelands. Our risk was only 'extreme'.

Within weeks, that green tinge of perennial grasses had hayed off; the annuals had been dead for months. In mid-November the weather bureau predicted a return to 'normal' rainfall (whatever that now meant) by February 2020, meaning we faced one more summer – assuming the bureau was correct – without rain.

Once again, my father decided we would wean our calves early and buy calf kibbles to keep them alive. But this summer presented new challenges, the cumulative result of a drought that was now as old as a mature cow.

All Gollion's dams were dropping, but the one that contained the pump we used to prime the siphon in a higher dam, from which we irrigated the orchards, was of most concern. When its water level fell below the pump's suction line we installed a temporary pump in the dam immediately above it, allowing us to top up the dam directly.

To top up the top-up dam, we next installed another solar pump a few hundred metres to the west, in the paddock we called Scribbly Gum, and unspooled hundreds of metres of pipe to transport its water, Dad scratching a shallow trench in the rocky ground with a mattock and me following behind and half-burying the pipe with my boots. This dam was spring-fed and self-replenishing, full and healthy – there were lilies on it and frogs in it, which I discovered when I stripped down to my boxers and waded in to help install the pump.

We turned on the new pump and by the time we had driven over the hill to the end of its attached pipe, water was pulsing into the dam, just as we hoped. 'I could sit here for hours just watching it,' Dad marvelled. 'I might even bring a little chair next time.'

He'd need a parasol too. On 21 November it reached 40°C on the veranda, three degrees off the hottest temperature ever recorded at Gollion, ten months earlier.

And a facemask. By the end of spring it had already been the worst fire season for years in New South Wales, Queensland and South Australia. That's when the smoke arrived at Gollion, appearing one morning from fires burning on the coast and one closer, eighty kilometres away. Visibility dropped to a few hundred metres. I took to wearing a red highwayman's kerchief over my face.

On 6 December, bushfires threatened the coastal village my parents were preparing to move to at the end of 2020. They were already spending half their time there to take Dad's mind off the drought. And now fire threatened their house. The road in and out was closed. Mum told me Dad wanted to go there to fight the fire, 'but I talked him out of that.'

Fires burned to the north, west and south of Sydney. Smoke alarms went off *inside* buildings. Stepping outside, said one newspaper, was the equivalent of smoking thirty cigarettes. It quickly became normalised. People donned masks and kept going about their lives. With fires burning the length of the Great Dividing Range, the government reiterated its support for coalmining. Refugees in the form of black cockatoos, a predominantly coastal species, appeared at the farm. The prime minister took his family on holiday to Hawaii.

Dad calculated we had one month left of feed in the paddocks, and that if it didn't rain by then, we would buy hay to feed out or 'destock to zero'. A few months earlier we had both been deflated when Brian from grazing school brought his current class for a field trip and told me afterwards that we were overgrazing.

In retrospect, we should've sold all our livestock by 2020. That's what the state's best regenerative practitioners did, as soon as the Indian Ocean Dipole starting trending positive after the second half of 2016 turned out to be the wettest since 1989. The dipole, a climate phenomenon affecting ocean temperature in the Indian Ocean and rainfall in Africa and Australia, is becoming the most accurate tool for farmers to predict rainfall in southern Australia.

One of these farmers, Martin Royds of Jillamatong, the beef cattle property near Braidwood I'd visited with grazing school, will manage his soil and grasslands so well that when his local council runs out of water, Martin will offer to sell it some of his, still clean and flowing in his healthy creek. Royds anticipated: he was proactive; he paid attention to subtle shifts that augured bad times ahead, and de-stocked accordingly. But neither he nor anyone else knew exactly when it would rain again, and that was the crux.

I was now spending most of my farm days trying to keep trees alive. I irrigated my figs once a week instead of fortnightly. I hauled buckets and water containers to trees outside the orchards. A new tagasaste plantation Dad and I had planted in 2016 was dying; oak varieties I had selected on account of their drought hardiness – Algerian, cork, holm – were dying. I got up at 5.30 each morning to avoid the heat; as children we are taught not to look directly at the sun, but this summer you could: sometimes it looked like the red orb of an Asian megacity; sometimes like the pale yolk of a battery-farmed egg. The light it cast that bushfire summer was stained-glass eerie, bright as red cellophane. And then each evening, when the wind turned and the easterly blew off the coast, it brought with it new smoke. You could actually see the cool change coming in.

One evening in mid-December, I walked with Suey Dog to let the cattle into Rear End. Gollion's hills were now a lifeless grey, like cement. From the house we walked through Siphon, then Solar and the paddock called Tree of Heaven, recently subdivided into a new paddock called Billy's Coat Hanger for my nephew Billy, a daydreamer who eschewed toys and who had played with a coat hanger for hours while Dad and I constructed the fence across the paddock that would take its name.

I made a point of remembering what I saw, felt and heard on that walk; a kind of mental *before* to be filed away in case the drought ever broke and we became accustomed to the *after*. The easterly had picked up, ruffling Suey's fur as she trotted beside me. We walked across spear grass, usually ankle-high and the bane of woollen socks but now easily stepped on, crunchy as day-old snow. With every footfall came splashes of grasshoppers. I knew enough about farming now to know they would be worse in a month. Grasshoppers lay their eggs on bare ground.

I saw crows – fifteen in one stringybark tree – and a translucent snakeskin that made me jump. There were haggard kangaroos, forced east by even worse conditions in the west of the state. They prefer short grass to long, and in one paddock I counted a mob of thirty grazing right to the ground.

The creek no longer held water but was at least a mane of green amid all the grey. There must have been puddles somewhere, because a heron took flight when I approached, disturbed in its hunt for frogs.

When we reached Rear End the cows were not mooing but hooting, demanding to be moved instead of waiting. I opened the gate and the herd of fifty we still owned rushed into the new paddock, not yet aware that there wasn't much grass there either.

As the cattle moved past me, kicking up dust with their hooves, I watched two galahs flying east, straight into the wind. They rose as one and were swung from side to side until there was a lull and their flapping propelled them forward.

On the way home, Suey Dog's ears were pinned back by the headwind, but they stuck up with each kangaroo she saw. When we crested the hill in Kungsladen, the smoke, heavier than the air, hung in the valley far below us. Normally this was my favourite view on the farm, and I liked to watch the white specks of cockatoos flying between dead gum trees. Today the view was the window seat of an airliner flying over clouds.

Records fell cheaply. Tuesday 17 December: the hottest average maximum temperature ever recorded in Australia (40.9°C). Wednesday 18 December: the *new* hottest average maximum temperature ever recorded in Australia (41.9°C). The next day was the hottest in Canberra in December (39.3°C); two days later it reached 41.1°C. One of those days I came inside after watering trees and lay down on the slate tiles. Mum said I was slurring my words. She said I had heatstroke.

Lauren was staying at Gollion, along with her brother, Paddy, and our friend Little Soph. It was too hot for them to go outside. You couldn't live here with kids, said Lauren. Her tone was accusatory. It was fair – she had only seen Gollion in extreme drought. As the summer wore on, there was tension between us. I didn't know this was what you meant, she said, when you wanted us to spend the summer together. There was respite on the horizon for us both. Lauren had been invited to a writing residency in Paris from March 2020, and my parents had agreed to take care of Gollion while we relocated to France for three months.

The night before, a massive swathe of Gippsland had been evacuated; today, new fires exploded across New South Wales and Victoria, closing roads and stranding holidaymakers.

At Gollion, I woke at 6 a.m. on 31 December, put on the fig drippers and checked the cattle. By 8 a.m. it was hot; by 10 a.m. a westerly was roaring – what Dad called the bushfire wind. At midday the cross-wires of my fig orchard snapped; I fixed them in what felt like a smoke blizzard.

I walked to the cattle to check they had water. It was 41°C degrees as I passed the veranda thermometer. Fires now burned from Tasmania to Queensland, but there was so little grass at Gollion that if a fire started here it wouldn't burn for long. Normally in summer we modified the grazing regime to account for fire danger, heavily grazing the paddocks surrounding the house in the weeks around Christmas. Now there was little to graze. I had to hold onto my hat and look down as I walked. The ground scuttled with gum leaves; a cowpat, dry as plywood, cartwheeled past me. The only sound was the wind in the trees. When I heard tweets I was surprised to spot two small finches, sheltered in the lee of a scribbly gum.

We weren't in the mood for a party, so Lauren and Paddy instead made a tagine from Gollion lamb and Gollion apricots. As we ate, rain drops – faint but unimagined – started hitting the tin roof. 'What's this stuff?' joked Dad. If it turned out to be a measurable amount, he reminded me, it wouldn't count in 2019's rainfall, because it would be recorded at 9 a.m. on 1 January. At 9 a.m. in the new year I emptied the rain gauge. It didn't contain a measurable amount.

It was reported that smoke from Australia's fires was turning New Zealand's glaciers brown. The next day, Canberra had the worst air quality in the world.

I was in a pub in Sydney that Saturday when my friend Harry called. He said he was 'in a daze', walking around his own farm, wondering what would burn. Harry was 'extremely concerned' about the next day's fire danger. I wasn't aware the forecast was any worse than it had been for months. That's what the politicians want, said Harry – we can't let them normalise climate change. He then asked me if we had a tractor at Gollion.

What did Harry want, asked Lauren. I told her. She logged in to Twitter:

'The farmers of Yass Valley are talking about a convoy of tractors all the way to parliament house.'

I got a text from a journo mate almost instantly: 'When? How?'

It didn't happen, not least because I was still learning to drive Gollion's tractor through Gollion's gates without taking Gollion's gate posts with me. I thought it was a fun idea – like José Bové, I told Harry, the French sheep farmer and anti-globalisation activist who in 1999 destroyed an under-construction McDonald's franchise with his tractor. But I also didn't think Harry was serious. I wasn't yet angry like he was.

I wasn't angry later that day when I heard that Penrith, west of Sydney, had recorded a maximum of 48.9°C, making it the hottest place on Earth. (The highest temperature ever recorded in Australia, officially, was 50.7°C in Oodnadatta, South Australia, in 1960. Oodnadatta is in the centre of the continent, in the middle of the desert. Penrith is fifty kilometres from the beach. Clinging to the coast is no longer saving Australians.)

Nor was I angry when Mum texted me that night, saying it had reached 43°C at Gollion and that their house on the coast was again threatened by bushfire – I wasn't angry that my parents might have to build a new house for themselves in their seventies, and I wasn't angry about what this could mean for the succession of Gollion.

I wasn't even angry when, again and again, the prime minister and state premier refused to say whether the fires were being exacerbated by climate change. Australian politicians representing the interests of the fossil fuel industry: hardly a revelation.

My anger surfaced one morning unbidden, the culmination of months – years – of torturous repetition, the same media story looped, on the TV, the radio and (especially) in *The Land*, which Dad bought each Thursday and plonked down on the kitchen table beside his Aldi catalogue.

What finally tipped me over the edge was a radio interview with a farmer in Queensland. Through tears she outlined how hard the drought was hitting: the heat, the water-and-feed stress, the smoke. She and her husband had kept all their cattle alive, but it was getting harder, she said. Through long pauses and heavy breathing, she told the presenter, a high-profile political commentator, that her husband was spending that day in his tractor, pushing over trees and scrub to feed his cattle. I waited for the presenter to ask the farmer why her husband was destroying the surviving plants on his farm instead of selling the cattle. This wasn't like previous droughts – improvements in transport and communication, coupled with a booming demand for Australian beef, meant you could always find a buyer.

Instead, the presenter said that she was sure she spoke for all her listeners in conveying her thoughts to the woman and her husband during this tough time.

'You are FUCKING KIDDING me,' I hissed at the kitchen radio.

You often hear Australian farmers maintain that *they* are the real environmentalists, not the 'greenies' in the cities, sipping their rhetorical lattes, as if farmers don't chug coffee by the mug, and as if cities are somehow outside the environment. In one of the most urbanised societies on Earth, that's how it's usually framed: the bush vs the city; the ideologues who venerate nature because they can afford to, pitted against the pragmatists who need to make a living off it.

I have sympathy for conventional farmers who think environmentalism is a luxury. They may feel trapped by a lengthening global food system that demands of them cheap products produced fast year-round. Corners – environmental, mental, nutritional, social – will be cut if consumers expect milk to cost $2 a litre and strawberries available out of season. But my sympathy is waning in the face of growing evidence that regenerative farmers are both mentally happier and wealthier than conventional ones.

Even in this, the worst drought of Dad's time at Gollion, as his friends and neighbours went into hundreds of thousands of dollars in debt buying hay so they wouldn't have to destock – an option they considered a failure of agriculture and of character – he would end up making a modest profit. It would've been higher if we'd started destocking earlier.

Such farmers, it seems to me, don't see themselves as custodians of a landscape, but producers of a commodity. They are engaged in a kind of mining, of topsoil instead of copper or iron ore. When their deposit is exhausted, they move on or go bust. This is the view that pits caring for country and making a living in irreconcilable opposition.

The dichotomy, I think, has become entrenched in Australia because of the mythical power of the white farmer. In the prevailing cultural narrative, the farmer, like the Anzac soldier, is beyond reproach. He is a nation-builder who battles an 'unforgiving' land to put food on our tables. It is in the farmer that the 'real' Australian character can be found. That he was at the vanguard of frontier violence – displacing one culture so that another could settle – is rarely mentioned, and when it is, the genocide is tacitly justified by the conviction that since the arrival of 'civilisation', virgin soil has been made *productive*.

Perhaps this is why you meet 'third-generation' farmers but not third-generation undertakers or accountants, although they surely exist. Among farmers, it is a boast: I have clung on; I 'tamed' an unruly environment. I've not met Aboriginal people who call themselves 1000th-generation hunters – the land seems more forgiving if you belong to a culture that lived here just fine for upwards of 65,000 years.

Telling me of all the tax breaks and concessions primary producers in this country enjoy (my father included), the fuel rebates and drought payments, Dad – a free-trade economist before he farmed full-time – shook his head with embarrassment. That's what happens when the farmers are political kingmakers, he said. And so poor land management isn't criticised but rewarded. The media reporting of the drought never, to my knowledge, questioned the farming methods of those most stricken by it. The message was clear: farmers deserve sympathy, not accountability.

I thought of something Wally Bell said on camera in the short film made about Derrawa Dhaura's listing as an Aboriginal Place: 'We as Aboriginal people have a very strong connection to the land. It's our belief that we originate from the land, we're here for a short time and then eventually we return back to the land: it's all about that connection to country.'

How very different from the vision of the *man on the land*, his boot on its throat lest it rear up and attack.

Black Summer provoked a national conversation, but the threads weren't talking to each other. Conventional graziers wanted money so they could buy more hay and keep farming as they always had; conservationists wanted stricken farms to be turned into national parks. Irrigators wanted more water allocations to keep their crops alive; conservationists wanted that water redirected to river systems where fish died flapping in fetid trickles. We needed more 'hazard-reduction burning'; we needed a government that cared about climate change. We needed to stop building houses in the bush; we needed to accept they would burn. We needed more nature over here; we needed better farming over there. Farming was something you intensively managed. 'Nature' was something you left alone.

When friends and acquaintances asked about the impact of the drought on my family and my farm they were always well meaning but often naïve and judgemental. Now might be a good time to stop farming cattle entirely? Would I just plant native trees when I took over? Would we dig more dams? Would I rip out the willows from the creek? I realised that many people I knew held the view that farming was inherently destructive, and 'nature' was what happened if you stepped away. Domesticated species were pitted against wild ones. Natives were good; introduced, bad.

Sometimes the tensions played out in our own household. Mum loved gum trees and the native birds they attracted; Dad thought them 'bludgers' that drained the soil and made more work for his beloved introduced tagasaste trees, which pumped nitrogen into the soil. Mum called wattles 'cheerful' for their bright yellow farewell to winter; Dad grumbled that they dropped their limbs on his fence-lines. (When I took over Gollion, I would be careful to plant the wattles away from fence-lines and the eucalypts on the top of ridgelines, where they weren't in the best soil.)

One friend suggested, given their methane emissions and the impact of cloven hooves on Australia's soils (a qualm

I suspect originates in the damage sheep and cattle have done to the banks of Australia's waterways – damage that can be avoided by fencing off waterways or not overgrazing their banks), that we should 'return the paddocks to their natural state.' What was that, and when did it exist? In 1788, when white people first arrived, with their sheep and fences, their set-stocking and overgrazing? Or 3500 years before that, when dingoes became the apex predator of Ngambri, Ngunawal and Ngunnawal Country? Eight thousand years before that, when eucalyptus became the dominant flora? Further back, when firestick farming wasn't practised because recycling of grasses was done by massive herds of herbivorous megafauna? Humans were responsible for all these changes.

I was reminded of a night I spent, in the winter of 2015, with animal rights activists I was writing about who opposed the culling of kangaroos in Canberra Nature Park. The day before, the head of ACT Parks and Conservation had told me the cull was necessary to ease grazing pressure on biodiversity; he'd said no such culling was needed in nearby Namadgi National Park, because dingo populations there remained intact. The anti-cull activists dismissed this as 'playing God'. Nature, they said, 'always sorts itself out.'

Today there are more kangaroos in New South Wales than people; my family sometimes seeks permission from the government to shoot them because they damage fences and compete with stock for pasture.

In a confused legislative response to the duality of the kangaroo (vermin and native), we must attach special tags to each culled animal, but we cannot use the carcasses 'for a secondary purpose'. If I want to eat kangaroo legally, my best option is to drive twenty minutes to the nearest supermarket and buy a steak from an animal shot perhaps thousands of kilometres away, packaged in Adelaide and sent to Canberra.

We can't return to the past, but we can shape the future. To live in and of this landscape under climate change will require an understanding of how this landscape once worked, and how it has stopped working. We need to manage the land in a way that will feed 25 million people (and counting), build soil, retain biodiversity, capture carbon and revitalise Indigenous connection to country. Too much weight is currently placed on the *what* and not enough on the *how*; the set and the characters, rather than the plot and the playwright – us.

Rewilding can mean re-introducing existing species into an ecosystem in which they are no longer present, but it can also mean using a living species as a surrogate for an extinct one. Introducing species – cane toads, foxes – has more often than not been catastrophic for Australia, but if done carefully it can mimic what has been lost.

Timed grazing of cattle can achieve a similar effect on pastures as firesticks and megafauna once did. Fossilised pollen in the central Australian desert suggests the extinction of grazing megafauna precipitated the collapse of a complex grassland. Could we recreate that through the careful application of cattle?

Indigenous land management has been refined over thousands of years. Following the arrival of Aboriginal people in Australia, there was a mass extinction of megafauna, followed by 60,000+ years of relative stability. In other words, it took people a while to realise how to live here. They couldn't do as much damage as we can (small population, fire as the only tool), but they still did damage. We don't have the time they were afforded, but we can tap into the lessons they learned.

During the Black Summer, the federal government called for a future increase in Aboriginal cultural burning to reduce bushfire fuel load. But they didn't call for any other measures that might help both grow food and mitigate the effects of climate change.

In the aftermath of the 2009 Black Saturday bushfires, for example, anecdotal evidence emerged that biodynamic farms didn't burn. Why not? Perhaps because plants grown biodynamically are more mineral-dense. It's why they're harder to digest – and why you need to eat less of them. Anything that is mineral-dense is not as flammable. Healthy soils would alleviate fires (there is evidence that more hydrated soils don't burn) as well as flooding (there is less run-off in country with good groundcover). Water can be retained where grass has returned. Grass can be returned through the manipulation of grazing animals. A healthier soil also means the plants are greener; they retain more moisture and are less likely to burn. And a healthy aerated soil, even in drought, will attract dew, which can be enough to stop it from catching fire.

Some of the firefighters who battled the blazes of Black Summer described trees collapsing in a way they hadn't seen previously. Was it because trees are less mineral-dense than they used to be? Because the soil they grow in isn't as healthy? Because the whole system is unhealthy? One compared it to a weak immune system catching things we normally wouldn't catch; we are 'catching' bushfires we normally wouldn't catch.

Martin Royds, having destocked before the fires arrived in his district, saw that the flames skimmed across his property because he still had groundcover, but on neighbouring properties the soil caught fire. Dead root matter with no soil microbes to eat it becomes as flammable as cardboard; it catches fire *underground*. If we want to address fire risk, we need to pay attention to what's happening under our feet.

I heard none of this from the government that summer. Instead, they spoke of digging more dams, even though a healthy landscape stores far more water. As whole towns began to run dry, the prime minister urged Australians to pray for rain.

Harold the 'carter' backed his Isuzu up to the rickety wooden loading ramp at Gollion's cattle yards and raised his pneumatic tyres with a hiss so that the bed of his truck was level with the end of the ramp.

Climbing down from the cab, he pulled on his black Akubra with thick sausage fingers and fetched his cattle prod from the passenger seat. Harold wore black stubbies, a blue work shirt and thick navy socks that ended in lace-up boots. He called the cows 'girls' and the weather by the third person singular feminine.

'Well, David, she's dry, isn't she?'

Harold had come to Gollion three times already that January. Sometimes we squeezed thirteen cows onto his truck; sometimes only ten would fit. Finally we were alleviating the grazing pressure on Gollion. My job was the gate-opener, sorting the stock into lots according to Harold's requests based on the available space in the pens of his truck. ('THREE cows and ONE bull, thanks, Sam'; 'FOUR small cows'). They would spend the night in Yass and be sold the next morning. The size of Australia's cattle herd, Dad said, was plummeting.

Once I'd sorted them into lots, Harold walked each animal along the race, talking to them all the way onto his truck. ('Walk up. Walk up. Carn', girls, walk up.') If they stopped moving he whistled like a willy wagtail, and if that didn't work he poked them with his electric prod. The bulls waddled up the race in their own time no matter what Harold did. I was especially sad to see them go.

The day before each visit we mustered what remained of our herd into the yards to decide which would be 'put on the truck'. The cows we sold were old or dry or 'big-framed' (they ate the most). We would sell the skinniest animals last, said Dad, to give them a chance to fatten up. Usually my father took his time in the yards when drafting the stock, but now he was fast and loose with his selections, almost arbitrary. He didn't care anymore, he told me. 'We'll probably end up selling them all anyway.'

Mum later told me that Dad said he felt 'heartbroken' selling all these cows he'd bred up himself. Mum had consoled him that whoever bought them would think, 'Look at these lovely big cows, I'll buy them for breeding.' Dad told Mum he was sick of farming.

Harold's most recent visit left us with nineteen adult cattle in one paddock and thirty-eight weaners in another. We would sell the rest of the cows in a fortnight, said Dad, followed by the weaners as soon as they reached saleable weight.

'And that will be the end of our business.'

That eerie red cellophane light I had come to hate was back; the smoke was picking up on the easterly. I told Dad he was being melodramatic. We could hold out selling our heifer calves until last, and then if the season improved we could use them to breed. He pointed out that would mean two years without an income stream, while I waited for those heifer calves to reach sexual maturity, and then waited for them to give birth to calves I could grow out to saleable wait. 'But that's for you to decide,' Dad said.

I felt guilty admitting it to myself, but I liked it when he spoke like this. When Dad made chitchat with Harold – when Harold first got out of the truck, and while they filled out the paperwork on the bonnet of the ute after the cattle were loaded – I held back, the diligent farmhand. I was always quiet, mostly because I felt I didn't really know what I was talking about. But I was privately chuffed when I overheard Dad say, 'I've had a gutful of this. I want to sell them all. Sam can build the herd back up – he's taking over here.'

Satisfaction turned to angst when I wondered what exactly I was taking over. No rain was yet forecast. But when it did start raining again, as it surely must (*surely*?), what future was there for agriculture in this part of Australia? When I asked Dad what he thought the climate would be like here in years to come, he said, 'Wouldn't have a clue.' He wasn't even pretending to be in control anymore.

Usually when Dad didn't have a clue about something, he wasn't so honest. When Harold had been trying earlier that day, with great difficulty, to back his truck up to the Gollion cattle ramp, he had said, 'I've gotta let the air out of the bags.' At first Dad smiled and nodded along like he understood, but then when Harold said it again (the truck engine was going, so it was hard to hear), Dad replied, 'How do I do that?' (He thought Harold was asking him to do it.) Then when Harold said, 'No, I've gotta let the air out of the bags', Dad said: 'I got fitted for hearing aids yesterday, but I don't want to wear them.' Harold smiled. He then tried to back the truck up again, and again failed. 'I'll just let the air out of the bags,' Harold called out the driver-side window, as if he thought of that just now. 'Righto,' said Dad. 'What bags?' I asked Dad so Harold couldn't hear. 'No idea,' said Dad.

The forecast for the next three days: 37°C, 41°C, 41°C. When Harold left and we were driving back up the lane, Dad said: 'I'd get your figs watered before this heatwave hits if I were you.' Even though he thought we might run out of water by the end of February, Dad advised against rationing what water we had, because then the fruit would be smaller than usual. Instead he told me to 'go for broke' and hope for rainfall to finish off the ripening. More dams were drying up completely every day, their bottoms cracked and crisscrossed with bird tracks. Mum had already sacrificed much of her garden.

That evening I walked to the creek. The air quality was good, with no smoke. The light was dark blue, the trees to the west silhouetted by the last light of day. I followed the ridgeline; from up here, to my north-east, I saw the lights of Gundaroo. Due north, a long way north, I watched lightning strike. I turned to the west and felt the wind at my back. In the sky I found Jupiter, strong and bright. I was flushed with a strong sense of belonging. On the walk back to the house I couldn't see but I didn't need to. I let my other senses do the work and navigated by familiarity. I *felt* this landscape, these hills, this ridge. Occasionally I disturbed a kangaroo. I clapped as I approached so they wouldn't accidentally hit me. The drought had rendered the farm unrecognisable, but it had never felt more familiar.

I was in Perth visiting my sister Sarah when the drought broke. The rain fell on Sydney first, my phone telling me the city had passed 50, then 100, then 150 millimetres since 9 a.m. on 9 February. Canberra stayed stuck on 0 millimetres. I became despondent: it wouldn't have been as hard to bear if heavy falls hadn't been forecast for Canberra too, but they had been. It wouldn't have been as hard if other drought-stricken parts of the country weren't being inundated, but they were.

Mum and Dad weren't at Gollion either, and were receiving text updates from their neighbours, Nick and Ali, which Mum passed on to me. Ali had 'confirmed their [our] worst fears': it still wasn't raining at the farm, although Ali thought it was 'trying' to rain. I knew what she meant: so many times during this drought, hearing rain start briefly on the tin roof, only for it to ease off again, I had found myself willing it to keep going, while also pretending it wasn't easing off: I'd hear rain on the roof when it was no longer there to hear.

I started checking the weather bureau website on the hour. Sarah cooked dinner, and, at 7 p.m., I left the table without explanation to grab my phone. It had started raining in Canberra. Earlier in the day, people online marvelled at the seeming inability of this 'east-coast low' – Sydney's biggest rain since 1998 – to reach Canberra. But at 8 p.m. it was still raining in Canberra and at 9 p.m. too. We passed Darwin's rainfall of 7.2 millimetres – a benchmark, however modest – and I went to bed feeling optimistic.

I woke to good news: Mum texting: 'Nick says first time our road has flooded since Lucy's wedding in 2010!'

I got into Canberra at 11:30 p.m., but I still had to ride my bicycle forty-five minutes to where I had left my ute. It was no longer raining, but the roads were wet with slick. The city's parks had a dark-brown, sodden look. In a peri-urban industrial complex I came across kangaroos; I nearly hit one when my light's battery failed. They would be hungry the next few weeks – even hungrier than usual. What little remaining dry feed there was had washed away.

In my car and out of the city, the moon cast its silver sheen on now-full dams and sheets of water puddled beside the road. At the little bridge over Spring Flat Creek, Gollion still five minutes away, I turned off the engine and listened to the frogs. The creek was flowing underneath my car; the creek was full. Passing Fernleigh, I could see that empty dams we had paid a neighbour to clean out and deepen with his bulldozer less than a week earlier were now overflowing.

Mum and Dad had arrived home that afternoon. I turned on the kitchen light and checked the rainfall chart, pinned to the pantry cupboard. Dad's scrawl was bold, rushed, excited: '**83 mm**'. I had made a recording of the frogs on my phone and, not knowing how else to celebrate, I listened to it in bed until I fell asleep.

# 17

The drought was over. Dad would need to find something else to complain about.

Lauren suggested a global pandemic – could he complain about COVID-19 like the rest of us? Her dream of living in Paris had been postponed indefinitely; we were due to fly out a few days after she called me in tears to say the border was closed. Instead of three months in the 4th arrondissement, my girlfriend faced a year of patchy 4G in my parents' farmhouse, parents *inclus*.

But I am yet to hear my father say a bad word about the novel coronavirus. His pre-pandemic lifestyle made him ideally suited to lockdown: social distancing, sheltering-in-place, living in a household where one person was assigned to do the grocery shopping and that one person wasn't him – he'd been observing them all for years. He even had some Ivermectin in the shed somewhere, he said with a wink, left over from an infestation of cattle lice years earlier.

At the nadir of Black Summer, the effects of drought on the land had us questioning whether *anything* would grow in our soil again, but the doubts proved temporary.

On 13 February it rained again, and a panel of corrugated Perspex came loose on the tool-shed roof. In a full-length Driza-Bone and a soggy Akubra that crinkled my vision, I crouched next to Dad and helped him replace the panel with a sheet of iron. Dad held it in place while I used a punch to make indentations for the screws, then drilled each one 'home' until Dad yelled, 'THAT'LL DO!' When we had finished I carefully stood up and noticed a green tinge on the hills where clover had sprouted after the drought-breaking rain. Today was what farmers call the follow-up.

Soon those hills were so deep with clover they were deemed *too* lush for grazing, Dad explaining that rapidly growing grasses, especially legumes like clover, can give cattle something called 'bloat', where foam forms in the rumen, preventing belching and, if left untreated, causing death.

There was a follow-up to the follow-up, and for several happy days we squelched around in gumboots together, checking on the newly cleaned-out dams and the drainage lines between paddocks. 'A serious farmer,' Dad said, 'would stake all this out and come back with trees to plant.' (I took photos instead.) He lamented that there wasn't more grass to stop all this water washing away, and within weeks he was lamenting that we'd sold so many cattle.

The catalogue, printed on glossy cardboard in funereal golds and blues, arrived in the mail that July. Dad flipped through it after dinner one night and tossed it across the table without having stopped on any one page. 'Here you go, Sam. Some homework before you buy your bull.'

I was buying a bull? I looked at him for an answer but he was already ogling a cordless drill for $49.99.

It could've been a brochure for an exclusive private boy's school, trying valiantly to reassure parents of prospective boarders that hazing had been against school policy for twelve proud months now. It sounded posh:

**HAZELDEAN**

**LITCHFIELD**

**EST. 1865**

Below a stylised 'H' (heads of wheat ripened on each corner) there was a photo of two bulls in profile – matte like liquorice, squat like cannons – both transfixed by something out of shot. What were they staring at? They were staring at cows. Cows waiting to be served. There would be an auction of ninety-five 'Performance Bulls', the catalogue said, at the stud on 3 September.

Most of the pages were taken up with arcane statistics on each of the bulls for sale. The catalogue was also full of diagrams – 'CLAW SET', 'SHEATH', 'REAR LEG HIND VIEW', 'REAR LEG SIDE VIEW' – and a scale of 1 through 9 indicating the desired anatomical structure of each. There was a lot of fine print: indemnity against a bull that *doesn't work*; indemnity against a prospective bull buyer being trampled to death. I was particularly taken by the 'Hazeldean Semen Terms & Conditions', which stipulated that buyers of Hazeldean bulls may only use their new stud to 'naturally service stock owned by the buyer' or to 'collect semen for use in stock owned by the buyer.' Breach Semen T&C and Hazeldean reserved the right to forcibly retake possession of the bull.

Dad told me he usually looked for 'curve benders': bulls that sired calves with a low birthweight (which makes calving easier for the mother) but which exploded in growth 'once they hit the ground'. As we sold most of our calves the autumn after they were born, the key growth date was 200 days – at 400 and 600 days they were 'some other bloke's problem.'

But that was in the past. The drought had so reduced our herd that we likely wouldn't sell any heifer calves for years, instead holding onto them as 'breeders'. That meant I should try to buy a bull that resulted in easy births, but also fast-growing calves that became big-framed adults. The bull I bought would also need to be suitable for joining with heifers – it couldn't be too big.

I read on:

> *Looking after your new Hazeldean bull*
> Bulls are like us. Some settle easily into a new location – others don't. Here are a few pointers to help your new bull/s settle in.

There followed a kind of parenting advice: how to avoid the new kid being picked on, and how to find him a paddock buddy for his first few weeks. Then this:

> Prior to mating a breeding soundness evaluation of all the bulls you own is a good idea. You may wish to undertake this yourself or get a vet to do it for you. Bulls should be assessed for reproductive soundness including palpation and measuring of the testicles. Testes should be at least 32 cm in circumference and tennis ball firm, not hard or soft. They should be symmetrical and free

> from abnormal lumps and bumps. The epididymis at the base of the testicle should not be enlarged or irregular and the cords at the top of the testicle also free of lumps.

I wondered if cattlemen checked themselves for lumps and tennis-ball firmness.

> It is a good idea to observe your bull's actual physical prowess in a serving ability test. When a bull is mounting a cow he must thrust and complete a service fully before he can be given a tick. Stand to the right and to the rear of the bull and check for corkscrew penis. Corkscrew penis is a condition where the end of the penis spirals, preventing entry into the vagina. Corkscrew penis will always deviate to the right side of the cow.

Pulling up a chair and watching the solar pump pulse out a trickle of water suddenly seemed so very soft-core.

The eve of the Hazeldean bull sale was the first spring night I could feel the warmth of the sun on the Earth after it had set. For a month now, Dad or I had walked around the herd twice daily, 'looking for legs sticking out of bums.' We may have needed a bull for spring joining, but the bulls we had sold in the drought lived on in the calves being born every day.

That afternoon I was checking the cattle at dusk. Sulphur-crested cockatoos, my favourite birds, raucously settled in the branches of a dead gum tree in the paddock we call Coleman East. That was where the cattle would be moved tomorrow, and they ran alongside me as I walked among them. The grass in their paddock had been grazed low. A pink super moon rose to the east.

In the southern corner of Dead Horse paddock, where a cow was licking its newborn when I checked the herd in the morning, there was now a cow all alone, and when I reached her I saw her amniotic sac was outside her vulva. I waited ten minutes. Sat in the grass. Made a phone call. Sometimes she put her tail up like a pooping dog; she sat down, got up, sat down, got up.

I walked home, told Dad she was 'about to give birth.' He asked if her water had broken. No. 'So nothing's changed. She's been like that for an hour,' he said. We returned to the paddock in his good ute. I noticed he'd already put the calf-puller in the tray, rusted with births of springs past.

It was dark when we reached the paddock. I plugged a spotlight into the cigarette lighter and scanned the grass. The cattle thought *this* time we were there to move them, and they bucked and reared in front of the ute, their eyes green in the spotlight.

The cow hadn't moved. I got out and, holding the spotlight in one hand, its curly cord stretched straight, and a piece of poly-pipe in the other, I started to make her move. The sac swung as she walked. It hung lower than her udder, which bulged in readiness for a calf. Her water broke, and now the sac looked like a safety parachute behind one of those land-speed cars they test in the desert, after they have come to a stop.

'Keep your torch on her,' Dad said from the driver's window. 'You don't want to lose her.' I didn't think that was likely. Even in the dark, she was the one with a safety parachute. Even in the dark, I could see that she had blood around her vulva. We walked to the gate, Dad driving slowly behind us. In the darkness either side of his headlights I heard cattle crashing through fallen branches.

We walked the cow out the gate – shutting it quickly so the rest of the mob didn't follow. I walked her down the lane, the parachute dragging in the dust. Frogs called from dams either side of the lane. A boobook owl started up.

We put her in the cattle yards, but not yet in the crush. 'Let's let her calm down for half an hour and see if she'll have a push,' said Dad. When a calf is normal, he explained, it is usually born within half an hour of the water breaking. 'But look how *fat* this cow is. She might just be lazy.' There was too much grass, Dad grumbled on the drive home, the car beeping all the way back to the house because our seatbelts weren't on. The cattle, he said, now had it too easy.

The cow was in the crush and Dad was in the cow. Tongue pressed to his upper lip in concentration, he felt for a front hoof. He pulled out his arm: there was a noise like the final gurgle the whirlpool of a bath makes before it disappears. He went back in: this time he got hold of a hoof, and pulled.

The cow shat on Dad's hands. I stood behind him in the narrow race that led to the crush, shining a torch. His pants were more shit than pants.

He pulled the hoof out just enough that I could slip one of the chains around it, on the skinniest part of the shank, above the hoof.

'Now you get the other one out.'

We swapped places, Dad keeping tension on the first

hoof while I ducked under his arm to get in front, an awkward dance move from a school social. I shone the torch with my left hand and entered the cow with my right. Warm and wet, not much room to move. She shat on my hand too, but by then I didn't care. All that mattered was getting the calf out. I got hold of the other hoof, still inside, and pulled. It barely moved, then slid back inside when I released it.

'We need more than that to put the chain around it,' said Dad.

I went in again – I hadn't even rolled my sleeves up like TV vets do – and managed to pull the hoof out a fraction more.

'Get the other chain around it.'

The other chain: where was it?

'Shit, Sam!'

'You had it! Did you bring it into the crush?'

Dad did a sweep of the grass with his torchlight, all the while keeping tension on his hoof.

'For fuck's sake, Sa— oh wait, here it is.'

He picked up the chain, then skilfully manipulated it into a knot with his left hand. I made a mental note to ask him later to show me how to do that.

'Slide it on,' he said.

I slid the chain around the second hoof and tightened the loop. Now we could try to crank out the calf by clipping both chains to a ratchet. But first we tried to pull it out, tug-of-war style, each of us holding a chain with both hands, our feet planted shoulder-width apart.

'PULL! PULL!'

I was pulling as hard as I could, but still Dad kept yelling, 'PULL!' The chain bit into my hands. It wasn't budging.

We started ratcheting it out, a chain on each hoof. We got ten centimetres more out, but the ratchet became tangled in the cow's tail. 'AH, *SHIT*,' Dad said.

I yanked bits of tail out of the calf puller, but we still couldn't get the chain around the other hoof. Instead Dad cranked both chains onto one hoof. A head emerged – covered in a blue-grey swimming-cap of membrane – then the torso, then the whole calf plopped onto the ground. A purple tongue lolled out of its mouth. I thought it was dead. Then I saw its eyes were open, big and watery.

Dad pulled the swimming cap off its head. He stuck two fingers into its mouth to stop it choking on its tongue. The calf was silent and barely moving. Its lashes, I noticed, were fully formed. Shit was all over it, and slime. Dad started tapping its cheeks like a mobster warning a wise guy.

'Open the gate to let the cow out of the crush, will ya?'

The cow stepped out – and walked away from her calf, to the corner of the yard where there was a thicket of grass. I heard her start grazing. Did she know what was going on?

The calf was slumped over and slimy, wet and dark. It looked like a water sprite, a selkie. It was struggling to breathe. We unknotted the chains around its hooves, then, me on the front legs and Dad on the rear, lifted it over the wall like a sandbag, into the yard with its mother. Still no maternal response. It was a bad sign: I knew by now that cows can tell when something is wrong with their calf, and in such situations reject it.

The calf made no noise. Dad said this was unusual. We looked again into its mouth (he said sometimes newborn calves can choke on mucus), but it looked okay. A bubble of snot formed on its nostrils with each exhalation, which Dad cleared away with the back of his hand.

Dad wasn't sure it was breathing properly and kept putting his ear to the calf's nose. I knelt down in the grass and put my ear to its mouth: faint breaths. The calf's chest was rising and falling, but its head was flopped in the grass and it didn't look alert.

We retreated to the edge of the yard and waited. Five minutes passed. Maybe ten. The cow kept grazing, cobwebs of afterbirth hanging out of her.

But then she walked over to inspect the bundle of slime, first sniffing, then licking the calf with her coarse purple tongue. She nudged it; Dad said she was 'trying to get it going'. The cow started making a noise – a kind of gentle, croaky moo. It looked up at us: it wanted us to leave. We started walking home up the hill but hadn't closed the gate yet when the calf called back to its mother.

# 18

My mother wanted to know if I'd like to keep the sandwich press. The piano? The grandfather clock? And what about this dusty textbook Papa was given as a jackaroo in the 1930s? It was a bit outdated now, but it might come in handy 'one day'.

They moved out in December. Mum wanted to hire a removalist but was rebuffed by Dad ('rip-off merchants'); instead he began ferrying the contents of their old house the 160 kilometres to the new one, couches and dressers and bedside tables jumbled together in a game of ute-tray Tetris, all of it tied down with knots learned as a Boy Scout sixty years earlier.

My mother's job was to separate the trash from the treasure, a thankless task usually reserved for grieving adult children. She told me she never thought she'd end up spending so much of her life in the same house, accruing vignettes of domesticity year upon year. And now here she was, prevaricating over whether to bin the paintings my sister Lucy had done as a preschooler during her own, less-heralded Blue Period.

I wondered what happens when the outgoing president

shows the new one around the White House. How do you respect the history of a home and its previous occupants but remain free to leave your mark? There was no ceremonial handing-over of the keys: my parents knew I was never going to lock the house, and besides, theirs would be a gradual decampment.

One day that month I drove my father to the creek. Dad's shirt was so ripped he looked to have been mauled by a bunyip. Suey rode shotgun. We parked before we could see any water, the sward of neck-high phalaris too dense to keep driving. There were no kangaroos on Gollion anymore – the grass was too high even for them. It was a still day, warm and partly cloudy. The floodplain hadn't yet hayed off and was a lovely pea-green.

Dad told me to get my phone from the glovebox, to take photos. 'You might not see it as good as this again here. I certainly won't.' When we reached the creek bank, we found the leaky weirs of piled-up rocks working as intended, water pooling behind each one as it queued to slowly trickle through. Stands of cumbungi and phragmites – two species of reeds that filter the water and slow its movement through the creek – were tall and swaying, having nearly disappeared during the drought. I asked Dad if he remembered stopping the car on the drive to the coast when we were kids and making us dig out hunks of phragmites to transplant here, waddling across the highway with dripping shovels weighed down by stolen native flora.

'Do I ever – Doughboy Creek was the place.'

Earlier in the day I had shown him the dozen oaks I'd sprouted from acorns in the driveway paddock and recently protected with tree guards. 'That's superb, Sam,' he said.

'I hope I get to see them mature – this is an incentive to live long enough.'

We walked across the leaky weir he called 'Sam's Crossing' and into the grass on the other side, looking out for snakes. Suey bounded through the grass, identifiable only by the sound of her sneeze each time a grass seed went up her nose.

The paddock here was a few metres higher than the surface of the creek, but it was wet and boggy. The weirs now worked so well that the whole water table had risen, brought even higher through the process of wicking or 'capillary action', by which plants suck water up from their roots to their leaves.

'To think what this looked like fifteen years ago,' Dad marvelled. 'It was just a bare ditch.' Now we could see herons, hear frogs, feel the temperature drop as we neared the water.

We ducked under the fence we had built together years earlier to keep the cattle from fouling the creek. 'This used to be a bare, yellow bank,' Dad said. It was now covered with native grasses and sedges.

'If there's one thing I've learned from farming, Sam, it's that 95 per cent of it is about the rain that falls, and how you retain it once it hits the ground. That's what makes you dollars, not fancy tractors.'

The lessons were never going to end, I knew. I was one link of a long chain of farmers; Dad was another. Papa, my father's farming mentor, had been dead for nearly a decade, but even that morning Dad had mentioned him, reminding me that sheep can stand long spells without water, but 'Papa used to say cattle get cranky if they go more than a day or two without a drink.'

Before we left, I wanted to show Dad a new weir I had started to build. It was hard going, so we headed away from the high grass along the bank, only to realise that it extended one hundred metres either side of the water, until the appearance of browned-off wild oats indicated that the moisture content of the soil had changed.

I asked my father how all this made him feel. 'That I've done something worthwhile with my life.'

There was one last job we needed to do together. My parents had moved to this district in 1983, in the aftermath of a drought, and in 2020 were leaving it in the wake of another. When a drought breaks, the logic goes, you make hay to prepare for the next one.

'I know, I know,' Dad had said when I reminded him that the best regenerative farmers don't cut hay; they consider it a disincentive to destock and a break in the cycling of nutrients between mouth, plant and soil. 'I know – but I want you to think of it as an insurance policy.' It was a profoundly loving act: if my father doubted whether I could 'make a go' of farming (as I suspect at least part of him did), he was determined to defer my failure as long as he could.

I told myself that when the next drought happened, I would remember the lessons of the one just passed. Act decisively; destock early; cattle can be bought and sold with a phone call, but overgrazed pastures take years to recover. But now, looking over the new bales of hay, I was made nervous by the sight of what amounted to cow fast-food, each weighing 300 kilograms and, when rolled out, enough fodder for our herd for a couple of days.

Whenever Gollion was next in drought, it would be hard to resist feeding instead of destocking. I think we both knew that when that decision came, I might not have Dad's wisdom to draw upon.

I stood in the middle of the paddock we called Hay Upper, freshly mown, golden and stubbly. My father cut the engine of his tractor and told me it was time for lunch. And then he smiled: it was also time to go compliment fishing.

'I must have had a lot of energy when I was your age, Sam – I used to do this job all by myself, *can you believe*?'

I could believe. Carting the hay from the paddocks to the sheds was hard, sweaty work. A few hours in and my face was caked in grime and straw dust, but Dad seemed more energised than when we started. 'All farmers understand that this is the busiest time,' he said. 'It's when you have to work like a Trojan.'

Fernleigh's two best paddocks had been 'locked up' at the end of August. The grass was allowed to grow tall as corn through spring before being cut, cured and baled at the start of summer. Once the hay was ready, the race was on to put it away before it turned to mulch.

Dad drove the tractor, stabbing the bales front-on with his forks, lifting them off the ground, then scraping them onto the back of the ute. A further six bales were plonked onto the back of a trailer we'd borrowed from a friend, and then I would drive Dad, Suey and the bales back to one of two haysheds, where the bales would be rolled into the paddock and stacked by tractor at a later date. We kept the ute running the whole time. There were no slopes to roll-start it in this flat valley floor, and while I waited for Dad to load me up I walked to nearby plantations, pulling weeds from among tiny oak trees I'd planted from acorns earlier in the year.

At the house we flipped through Dad's farm records, a dog-eared exercise book. We found what we were looking for in 2006: that was the last time he had cut hay. Over 300 round bales. Back then, the book said, we had 154 cows. Even in a good year, I now knew, running that many cows and their calves left little margin for seasonal or climatic variation.

'For years I ran that many, and fed them hay every winter,' Dad said. 'Mind you, I only started making any money when we stopped running so many.'

You hear this a lot from farmers who have transitioned from industrial to regenerative techniques. Farming within the constraints of nature may produce fewer commodities, but without the need for expensive inputs, profits rise. By running his herd holistically, Dad had increased the amount of winter grass, ending the need for hay during the colder months.

Soon after lunch we were standing on the back of the ute, rolling the bales off in front of the shed, when Dad lost his balance, falling backwards into the grass with the outstretched arms of a stuffed grizzly bear. He said he was fine, but I knew that his head had missed the corner of the trailer by ten centimetres. A few weeks earlier I had been in my room at Gollion when I heard a loud crash outside the window; my father had tipped the ride-on lawnmower off a steep bank in the garden, and when I found him he was trapped underneath it. He'd had the same stunned, slow-motion quality he did now, and had said only, 'Give us a hand to get this mower upright, will you? It's leaking fuel.'

This time, I helped him to his feet and back into the ute. I couldn't keep putting him in these situations. It was the right time for me to take over.

Two hours later I was walking among the bales, waiting for Dad to load up the ute, when I noticed he was waving his arms like an air-traffic controller. He opened the door of the tractor and turned off the ignition. What now?

'The linkage is broken.'

The 'linkage' is the pneumatics that controls the front forks. If it wasn't working, we couldn't pick up the hay. The joystick controlling it had 'come apart' in Dad's hands.

Hours passed in fruitless tinkering with screwdrivers and spanners ('there's gotta be some trick to this'); the light began to fade. I stood on the dirt floor of the shed, beside the tractor, while Dad worked in the cab. All I could see of him were his short shorts, tanned legs, boots and socks now covered, like mine, with grass seeds. His back brace lay abandoned in the dirt; he had hurt his back in the fall, and at afternoon tea Mum had suggested I call a friend to help me put the hay away in the event my father couldn't. My eyes grew heavy and I struggled to concentrate – to understand what it was we were trying to do.

Later that night, I overheared my father talking to his elderly uncle Russell on the phone ('Just touching base because I'm not sending you a Chrissy card this year'). 'Sam's right on top of the cattle and he's got a couple hundred fruit trees to keep him busy,' Dad said. 'He does things differently to me, but that's the way it goes.'

After Dad hung up, I asked him what he'd meant.

'Well, you're much more motivated to walk around the place, paying attention. That's okay – Papa was like that. The best farmers are like that.'

He paused. 'I also didn't want to say, but you're not a natural mechanic.'

Eventually Dad would call a mechanic, who worked out what was wrong through trial and error. ('That's a good lesson for you – he just fiddled his way through it. I've watched you – you don't do that.')

But at the end of that first day hay carting, in the tractor shed, my father had not conceded defeat just yet. The sun was setting, and cockatoos were noisily eating the blue-green wattle pods in a plantation behind us. Below them, two heifers rubbed against an old white Peugeot that had been relegated from the carport to the paddock once its registration lapsed. In the eaves of the shed, above us, a pair of swallows flitted about their nest, shrieking each time Dad dropped a spanner into the dirt with a clang.

'If only we could separate this case,' said Dad. Of what, I didn't know.

'Can we?'

'No idea. That's why we need a mechanic – someone who actually knows what they're doing.'

But of course, he kept tinkering, pulling the 'case' apart, handing me pieces, telling me I'd be 'in trouble' if I lost this bolt or that washer, scolding me for using my iPhone to help him see what he was doing ('You need a proper torch – you're living on a farm now!'). He hit his head on the corner of the tractor door as he climbed back into the cab ('I can't see a BLOODY THING with this hat on') – and then hit it again immediately. My father. I told him this was hopeless, that we should knock off for the night and return in the morning, but he asked me to give him a minute. There was just one last thing he wanted to try.

'Tell me,' said Dad, 'will *I* be in this book you're writing?'

It was a fair question. What were all those notes I'd been taking these past seven years, the destination of the answers he'd given to my dumb questions, asked atop orchard ladders and among the dung and dust of the paddocks? My father knew I was writing a book, but I guess he'd assumed it was some kind of agricultural how-to.

I laughed but said nothing, kicking off his boots on the back veranda and waiting for him to kick off mine. He'd put the wrong pair on again. Our feet are the same size.

The date was 10 July 2021, and my father was back at Gollion. Seven months had passed since he and Mum moved out, and although they returned often to help me, this time was different. Today, he wanted a tour.

The handover had not been easy. I naively thought the hardest part of a farm's succession would be the farming. But it was the familial that was most fraught. Through most of 2020, Mum, Dad, Lauren and I lived together. The farmhouse is big, but not big enough for that. Lauren moved into Canberra in October; she wanted to live with me, she said, not with my parents.

I had hoped the new year would bring clearer boundaries. After my parents moved out, Lauren moved back in. I painted the walls and took furniture to the tip; I tried to make the house brighter, less cluttered, more *ours*. But the aftermath of my parents' visits saw us emptying abandoned cups of tea and shutting doors; mopping up footprints and turning off a blaring TV. It was as if my mum and dad had heard a tsunami siren and fled for higher ground.

'This will always be their home,' said Lauren. 'They built the house, they established the garden. You can't move out of a farm you've lived at for forty years and expect to make a clean break.' She was right. We selected a site where we will build a small house of our own. A home in the bush, atop the ridgeline of Kungsladen paddock: The best view on the farm.

Our tour had begun in my fig orchard. It was a cold day, still and overcast. We ducked under the irrigation lines and walked along each row, the grass grey and the trees bare, each one in need, said Dad, of pruning.

Over the summer and autumn, I had sold all the fruit I could spare to a growing list of cafes, bars, restaurants and a gelataria. In the eighteen months since the drought had broken, my trees had grown more than in the preceding five years. Dad seemed impressed: 'It's beginning to look like an orchard.'

He told me of the fruit he was growing down the coast that he hadn't been able to here on account of the frosts: avocados, mandarins, ruby red grapefruit. It pleased me how well he had taken to his new home: he said he had never had so many friends and was annoyed he'd missed that morning's ride with his mountain bike 'gang'. (Mum had joined *two* book clubs in their coastal community but was still a member of one in Gundaroo. She returned to the farm more often than Dad to see her friends; while she kept a wardrobe of clothes in the house, Dad thought a bucket in the tool shed sufficed, stuffed with work shorts, a belt and a grease-stained shirt.)

We continued walking down the hill to the cattle yards, where Dad pointed out a broken fence rail I'd promised to replace. My worst tendency as a farmer, I have discovered, is to put off fixing things. The ute got a flat tyre and stayed parked in its shed for months; when the chain on my preferred chainsaw broke, I used another chainsaw until its chain broke too, at which point I consulted YouTube to build a new chain, too embarrassed to (again) ask my father how it was done.

But I have also discovered I don't need all Dad's skills. Gollion can be a shared resource and, in return, a place where favours are paid. Our friend David keeps his bees at Gollion; he collects firewood, hunts and has erected a shooting target. In return, Lauren and I get honey, kangaroo meat and wood. Wally Bell returned to Derrawa Dhaura in the summer and his sister, Karen Denny, is planning an art exhibition there. I have an idea I want to run past Wally, a collaboration to start farming native foods at Gollion.

My sister Eve thinks we should grant more of Gollion to its traditional owners. Her partner, Shane, still gives the tails and pelts of kangaroos he shoots at the farm to Aunty Matilda House. Shane and Eve's son, my nephew Ned, has become a crack shot. I've asked him to teach me to shoot. The last time his family visited the farm I taught them the names of some native perennial grasses, and how I hoped to establish more through careful rotation of the cattle. Eve, an anthropology lecturer, invited me to address her students on the subject of regenerative agriculture.

Cattle work I prefer to do alone. Working in the yards without my father was strange at first, but empowering: I bought the herd from my parents in the autumn and set about rebuilding its size, weaning the calves in May and running them as a separate mob instead of customarily selling them in Yass. The females would be joined in the spring with my new bull; the steers, I told Dad, I would sell once they weighed 500 kilograms. But I will never run the same cattle numbers at Gollion as my father once did. When the next drought inevitably comes, I will destock quickly.

'I think you're grossly overestimating how much money you can make from this place,' Dad said.

I told him that apart from dairy, cereals and olive oil, Lauren and I were now largely self-sufficient. On summer nights we had dehydrated nectarines and apricots, apples and pears. I juiced apples for us in an old-fashioned wooden press, and canned seventy bottles of preserved fruit on the stove to eat through the winter.

A brilliant, creative cook, Lauren uses whatever I can bring in from outside. We have eaten dandan noodles with kangaroo mince; lamb neck cooked on the woodstove in our own preserved tomatoes; stocks made from our own roosters. I am getting better at growing vegetables, and in April we planted a crop of 120 garlic plants for our kitchen. Food is the way we enjoy Gollion together.

My father shook his head at the effort of it all. 'That's one thing about living on a farm, I suppose – you're not going to starve to death.'

We next squelched across Cattle Yard paddock, our boots leaving indentations in the grass. Since the drought broke, the rain had been plentiful. ('The place looks fantastic, Sam – it looks like an *English* farm.')

At the fence-line I held the barbed wire down with my boot so my father could climb through and into our destination, Willow paddock. As I climbed through after him he told me he had been reflecting on his university class, and how those who returned to their own family farms after graduating 'didn't do as well' as those who became agricultural policymakers or consultants – they weren't as financially successful *because* they had a farm to take over.

I didn't know why he was telling me this. It stung, and I told him – defensively, unnecessarily – that although I know he thinks I'm poor, I have never felt richer. I had no boss, no colleagues, no commute. Every day I did exactly what I wanted, and I returned to the house each night feeling satisfied, not stressed.

We would always need an off-farm income. Lauren was working as an arts writer and as the curator of a video art platform she founded. She had not planned to live on a farm, but the move was surprisingly good for her career. In one of her essays, for the magazine *Kill Your Darlings*, she wrote:

> For years I have self-subsidised my own involvement in the arts by working as a freelance journalist, casual academic, research assistant and copywriter. I worked across three sectors – arts, public media and academia – that have all bowed to funding neglect by embracing short-term contracts. Eventually, during a series of failures to relocate overseas, I met someone who grew up

and lives on a farm, and when the pandemic hit, I moved in with him. I began to exhale. On stolen, stricken land, the farm subsidy has allowed me to detach from gig life by granting me that most precious asset: time.

In Willow paddock, my father and I reached the main mob of cattle. They were joined over the summer with the new bull, and by my calculation would start calving in early August. 'I think you'll have calves before that,' said Dad, looking between two hind legs.

We walked slowly among the herd, Dad reminding me to look at the udders to determine how far off the first calves were. An expectant cow will 'bag up' a few weeks before she calves, and many of the udders we passed were brimming with milk. Dad said the calves weren't imminent (he thought the udders were still filling and not yet 'tight'), but to expect the first within weeks. On a dry cow, he reminded me, you can barely see the udder; the animal will be fat, but not in the way of a pregnancy, in which the calf bulges prominently on one side of its mother's body.

'She's not far off,' Dad said of one heifer.

'She's not far off,' he said of another. And another, and another, and another.

'You've done well, Sam.' I smiled. The son's desire for paternal approval: as old as these hills.

It was a good omen. We didn't know it yet, but in July, Lauren fell pregnant. Our daughter, Orlando, will be born in March, when the figs are at their peak.

How long she lives at Gollion will be determined by climate change, suburban encroachment, if and when my sisters want to sell up, and my commitment to honouring Lauren's goals as she has honoured mine ('God forbid I die in Australia,' she likes to say). And, it will be determined by Orly.

But when I picture myself as a parent, it is at Gollion. I see myself with a baby strapped to my chest, fixing fences and planting trees. I think of the love and sense of belonging I now hold for this place, and how I want my daughter to feel those too. I think of all that I want to teach her – plant names and bird calls, how water moves through the landscape, and why her grandfather turns over rocks in the paddock to 'see what the ants are doing.'

It is a privilege to manage a landscape, Dad taught me, but also a responsibility. I will instil this in my own child: we are only custodians of Gollion for a short time, so while we are, we must treat the role with respect.

My father adjusted his beanie and we made for the house. He was cold, he said; he was already used to the balmy climate of the coast. We reached the veranda and he asked me about the book I'd been writing each night once my farm work was done for the day – would he be in it?

On every page, I told him.

He nodded slowly. He was thinking.

'Bloody hell.'

# ACKNOWLEDGEMENTS

Having a writer in the family must be hard enough, but spare a thought for my parents – they've got two. In 2006, my sister Eve Vincent wrote an essay for *Griffith Review*'s Gen X-themed edition deconstructing the generational divide between her 'ratbag', World Economic Forum–blockading self and our free marketeer of a father, set to the backdrop of the farm they both loved. So in 2017, when I was asked if I'd like to contribute to *Griffith Review*'s millennials edition, I decided to write a kind of a companion piece. It turned into this book.

Thanks, Eve, for shining your torch, and thanks to those members of my family who aren't writers but are written about here. Lucy and Sarah Vincent fact-checked several anecdotes and gave me plenty more; I especially love that your friends used to call Dad 'Groundskeeper Willie' on account of him being 'ripped for an old bald guy'. My mum, Jane Vincent, provided many loving and helpful suggestions, even if I didn't implement them all. ('Too many Aldi jokes!'). And of course, my father, David Vincent, who I hope isn't too embarrassed when he eventually reads it. I've so loved getting closer to you these past few years.

## ACKNOWLEDGEMENTS

This book would not have been possible without the guidance, support and tough love of Chris Feik. I'm eternally grateful you plucked my first book from the slush pile and invited me to join your stable. Also at Black Inc., Julia Carlomagno's enthusiasm helped me almost submit the manuscript on time, and once I had, Denise O'Dea beat it into shape. Thanks for your clever edits, Denise, and for your city-slicker sensibility each time I descended into barnyard esoterica.

My other early readers, Billy Griffiths, Alison Elvin and Joe Bartlett-Marques, made the book better in ways I hadn't anticipated, and Sari Braithwaite helped shape the structure after reading an early chapter.

Lauren Carroll Harris, my fearless samurai of a girlfriend, put up with a partner who went from showering weekly to fortnightly, and she moved to the country (in a pandemic, with the in-laws) so he could pursue his pastoral dreams. I won't let you die in Australia, honey, I promise.

Finally, *My Father and Other Animals* was bookended by hospital admissions. Six years before I started writing it, Dad's run-in with the woodchipper set the book in motion, and eight years later, just as I thought it was nearly finished, baby Orlando arrived. Fact-checking a manuscript with a newborn makes me realise just how few nappies the big boys of the Western canon must have changed. Throw a bumper fig season into the equation and for a few weeks there I was going to sleep at 4 a.m. and rising a few hours later. I have never been tireder, or happier. Already in your short life, Orly, we've explored much of the farm as a family, discovered a wedge-tailed eagle nest, 'picked' figs, made compost and helped a turtle cross the road. We're going to have so much fun.